T0290078

Accessing the E-book edition

Using the VitalSource® ebook

Access to the VitalBook™ ebook accompanying this book is via VitalSource® Bookshelf — an ebook reader which allows you to make and share notes and highlights on your ebooks and search across all of the ebooks that you hold on your VitalSource Bookshelf. You can access the ebook online or offline on your smartphone, tablet or PC/Mac and your notes and highlights will automatically stay in sync no matter where you make them.

1. **Create a VitalSource Bookshelf account at** *https://online.vitalsource.com/user/new* or log into your existing account if you already have one.

2. **Redeem the code provided in the panel below to get online access to the ebook.** Log in to Bookshelf and click the **Account** menu at the top right of the screen. Select **Redeem** and enter the redemption code shown on the scratch-off panel below in the **Code To Redeem** box. Press **Redeem**. Once the code has been redeemed your ebook will download and appear in your library.

DOWNLOAD AND READ OFFLINE

To use your ebook offline, download BookShelf to your PC, Mac, iOS device, Android device or Kindle Fire, and log in to your Bookshelf account to access your ebook:

On your PC/Mac

Go to *http://bookshelf.vitalsource.com/* and follow the instructions to download the free **VitalSource Bookshelf** app to your PC or Mac and log into your Bookshelf account.

On your iPhone/iPod Touch/iPad

Download the free **VitalSource Bookshelf** App available via the iTunes App Store and log into your Bookshelf account. You can find more information at *https://support. vitalsource.com/hc/en-us/categories/200134217-Bookshelf-for-iOS*

On your Android™ smartphone or tablet

Download the free **VitalSource Bookshelf** App available via Google Play and log into your Bookshelf account. You can find more information at *https://support.vitalsource.com/ hc/en-us/categories/200139976-Bookshelf-for-Android-and-Kindle-Fire*

On your Kindle Fire

Download the free **VitalSource Bookshelf** App available from Amazon and log into your Bookshelf account. You can find more information at *https://support.vitalsource.com/ hc/en-us/categories/200139976-Bookshelf-for-Android-and-Kindle-Fire*

N.B. The code in the scratch-off panel can only be used once. When you have created a Bookshelf account and redeemed the code you will be able to access the ebook online or offline on your smartphone, tablet or PC/Mac.

SUPPORT

If you have any questions about downloading Bookshelf, creating your account, or accessing and using your ebook edition, please visit *http://support.vitalsource.com/*

INTRODUCTION TO
COMPUTATIONAL
LINEAR ALGEBRA

INTRODUCTION TO
COMPUTATIONAL
LINEAR ALGEBRA

Nabil Nassif

American University of Beirut
Lebanon

Jocelyne Erhel

INRIA, Rennes
France

Bernard Philippe

INRIA, Rennes
France

CRC Press
Taylor & Francis Group
Boca Raton   London   New York

CRC Press is an imprint of the
Taylor & Francis Group, an **informa** business

A CHAPMAN & HALL BOOK

MATLAB® is a trademark of The MathWorks, Inc. and is used with permission. The MathWorks does not warrant the accuracy of the text or exercises in this book. This book's use or discussion of MATLAB® software or related products does not constitute endorsement or sponsorship by The MathWorks of a particular pedagogical approach or particular use of the MATLAB® software.

CRC Press
Taylor & Francis Group
6000 Broken Sound Parkway NW, Suite 300
Boca Raton, FL 33487-2742

© 2015 by Taylor & Francis Group, LLC
CRC Press is an imprint of Taylor & Francis Group, an Informa business

No claim to original U.S. Government works

Printed on acid-free paper
Version Date: 20150409

International Standard Book Number-13: 978-1-4822-5869-1 (Pack - Book and Ebook)

Library of Congress Cataloging-in-Publication Data

Nassif, Nabil.
    Introduction to computational linear algebra / Nabil Nassif, Jocelyne Erhel, Bernard Philippe.
        pages cm
    "A CRC title."
    Includes bibliographical references and index.
    ISBN 978-1-4822-5869-1 (hardcover : alk. paper) 1. Algebras, Linear--Textbooks. I. Erhel, Jocelyne, 1956- II. Philippe, B. III. Title. IV. Title: Computational linear algebra.

QA184.2.N37 2015
512'.5--dc23
                                                2015011621

Visit the Taylor & Francis Web site at
http://www.taylorandfrancis.com

and the CRC Press Web site at
http://www.crcpress.com

To the dear and supporting members of our respective families:

Norma, Nabil-John and Nadim

Yves, Marion and Mélanie

Elisabeth, Isabelle, Hélène, Etienne and Afif

Contents

Preface

This work results from two decades of common academic experience shared by the authors in teaching, between 1996 and 2003, introductory and advanced material in computational linear algebra and its application to numerical solutions of partial and ordinary differential equations. During that period, the authors worked as a team in a Master's program on "Mathematical Modeling and Numerical Simulation" managed jointly in Beirut by French, Swiss and Lebanese universities. Since 2003, that common experience has continued and is still pursued through multiple French-Lebanese collaborations in various research projects, teaching missions and co-tutoring of Master's and PhD theses.

The core of the book is adapted to a course on Numerical Linear Algebra offered yearly in the American University of Beirut to senior undergraduate students in Mathematics, Applied Mathematics and Engineering. Additional applications are also included. These are usually given to first-year graduate students in Engineering and Computational Science.

The main learning objectives of the book stand as follows:

1. In Chapter 1, the reader is exposed to BLAS operations of types 1, 2 and 3. These are particularly adapted to a scientific computer environment such as MATLAB® version 7. Please note that:
   MATLAB is a registered trademark of The MathWorks, Inc.
   For product information, please contact:
   The MathWorks, Inc.
   3 Apple Hill Drive
   Natick, MA 01760-20098 USA
   Tel: 508-647-7000
   Fax: 508-647-7001
   E-mail: info@mathworks.com
   Web: www.mathworks.com

2. Chapter 2 presents the basic mathematical tools needed in Numerical Linear Algebra: ranks, determinants, eigenvalues, vector and matrix norms, orthogonal matrices, Gram-Schmidt process, Schurs Decomposition and Singular Value Decomposition (SVD).

3. Chapter 3 gives the classical material on Gauss decompositions followed by LU and Choleskys factorizations of matrices. Additionally, it provides

the use of condition numbers for evaluating the effect of finite precision arithmetic when solving directly a system of linear equations $Ax = b$.

4. Chapter 4 illustrates the use of Householder transforms in obtaining the QR orthogonal factorization of a rectangular matrix that leads to finding its pseudo-inverse. This is followed by applications to least square solutions of rectangular systems and statistical regression analysis.

5. Chapter 5 is a detailed numerical treatment of the algebraic eigenvalue problem, starting with the power method and followed by the QR and Householder-Givens algorithms. Several applications are given as exercises, in particular examples from population dynamics and "Google" matrices.

6. Chapter 6 discusses at length (indirect) iterative methods to solve a system of linear equations $Ax = b$. It exposes stationary methods based on matrix splitting (Jacobi, Gauss-Seidel, SOR, SSOR) as well as Krylov spaces methods (steepest descent, Conjugate Gradient, GMRES and Bi-Conjugate Gradient). The determinant role of preconditioners is also exhibited.

7. Finally, Chapter 7 illustrates practices on solving discretized sparse systems of linear equations $AU = F$, obtained using either finite differences or finite elements when approximating the solutions of ordinary and partial differential equations. It provides a complete treatment of the problem from generating nodes and elements, computing local coefficients and "assembling" the sparse linear system. Various solvers are then implemented and compared in a number of computer projects.

The core material can be easily achieved in one-semester by covering:

- Sections 1.1 to 1.6 of Chapter 1.

- Chapter 2, without necessarily "insisting" on the proof of Shur's decomposition theorem.

- All of Chapter 3.

- Sections 4.1, 4.2, 4.4.1, 4.4.2, 4.5, 4.6 and 4.8 of Chapter 4.

- Sections 5.1.3, 5.2, 5.3 and 5.5 of Chapter 5.

- Sections 6.1, 6.2, 6.3 and 6.4 of Chapter 6.

The selection of additional topics, particularly that of applications, is left to the course instructor, particularly that regarding Sections 1.7 and 1.9 (sparse matrices and Strassen algorithm), Sections 4.11, 5.7 and 5.8, and selected material from Chapter 7 on sparse systems resulting from finite differences and finite element discretizations of ordinary and partial differential equations.
Throughout the book, special attention is given to algorithms' implementation using MATLAB's syntax. As a matter of fact, each of the numerical methods explained in any of the seven chapters is directly expressed either using a

pseudo-code or a detailed `MATLAB` program.

Exercises and Computer Projects
Each chapter ends with a large number of exercises. Throughout the seven chapters, several computer projects are proposed. These aim at testing the student's understanding of both the mathematics of numerical methods and the art of computer programming.

Nabil Nassif, Jocelyne Erhel and Bernard Philippe

About the Authors

Nabil Nassif received a Diplôme d'Ingénieur from the Ecole Centrale de Paris and earned a Master's degree in Applied Mathematics from Harvard University, followed by a PhD under the supervision of Professor Garrett Birkhoff. Since his graduation, Dr. Nassif has been affiliated with the Mathematics Department at the American University of Beirut, where he teaches and conducts research in the areas of mathematical modeling, numerical analysis and scientific computing. Professor Nassif has authored or co-authored about 50 publications in refereed journals and directed 12 PhD theses with an equal number of Master's theses. During his career, Professor Nassif has also held several regular and visiting teaching positions in France, Switzerland, the U.S. and Sweden.

Jocelyne Erhel is Senior Research Scientist and scientific leader of the Sage team at INRIA, in Rennes, France. She received her PhD from the University of Paris in 1982 and her Habilitation at the University of Rennes in 1992. She has been working for many years on parallel numerical algorithms. Her main subjects of interest are sparse linear algebra and high performance scientific computing applied to geophysics, mainly groundwater models. She coordinated and participated in several national and international grants and she published more than 90 papers.

Bernard Philippe was Senior Research Scientist at INRIA in Rennes, France, up until 2014 when he retired. After a 9-year period of teaching mathematics in secondary schools in Algeria and in France, he received his PhD at the University of Rennes in 1983 and his Habilitation at the University of Rennes in 1989. He has been working for many years on parallel numerical algorithms. His main subject of interest is matrix computing with a special emphasis on eigenvalue problems for large size matrices. He coordinated and participated in several national and international grants and published more than 60 papers. For three years he has also been Scientific Manager for the areas of Africa and the Middle East in the International Relations department of INRIA.

List of Figures

List of Tables

List of Algorithms

Chapter 1

Basic Linear Algebra Subprograms - BLAS

The library BLAS [24], [25], [44] defines a standard for the bricks on which efficient modern matrix computing is organized. In this chapter we define such types of operations and study levels of their complexity, both in terms of computer arithmetic and memory references involved in BLAS implementation. To use the full power of the processors, it is important to maintain a high value for the ratio between the number of floating operations, *flops*, and the number of memory references that measures communication time between memory and processor.

1.1   An Introductory Example

A simple computer architecture corresponding to a *Von-Neumann scalar machine* is one that is described by the following scheme

$$P(C, AL) \longleftrightarrow M$$

where M and and P are, respectively, the memory and processing units. The processing unit has two subunits:
The Control Unit C which "controls" the execution of program instructions according to a built-in "clock." Programs are stored in the computer memory M. Their sequential execution requires their proper interpretation by the Control Unit followed by their processing of data, also stored in M, within the Arithmetic and Logic Unit, AL.
On a processor, let us consider the execution of the following simple arithmetic instruction:

$$r = a * b + c/d \tag{1.1}$$

Assume the numbers a, b, c, d and r are floating point numbers stored in IEEE double precision format with $d \neq 0$. On a regular computer, (1.1) would require the execution of the following sequence of steps:

1. Move the contents of a, b from memory to registers in the arithmetic unit (registers are local fast memories in the computer processor). The way data is moved from memory is usually performed in several steps using intermediate memory such as caches.

2. Perform and save in register 1 the result of $a \star b$.

3. Move the contents of c, d from memory to arithmetic unit.

4. Perform and save in register 2 the result of c/d.

5. Add the data in register 1 to that in register 2.

6. Move back the end result of step 5 to memory location r.

Thus, the complexity of such operation can be evaluated in terms of:

1. **Transfer of data from memory to processor:** Such operations are in direct relation with the **number of references** that refer to the data stored in memory involved and consequently the **number of clock periods needed** for this data transfer to the processor.
   In the case of (1.1), there are 5 references involved: a, b, c, d and r. The transfer of data is enhanced by the number of "channels" between the memory and the processor and the speed of flow of data within the channels. Data can be transferred in a scalar way (one data at a time) or in a **pipelined** way. Modern computers are built in order to increase the efficiency of such functionality by two aspects: (i) the processor is composed of independent units which cooperate to perform arithmetic operations, (ii) there exits a memory hierarchy which allows to work on a bunch of data at the same time.
   Practically, since in (1.1), steps 1 and 3 can be performed simultaneously, the execution of (1.1) is reduced to:

   (a) Move the contents of a, b, c and d from memory to arithmetic unit.

   (b) Perform and save in register 1 the result of $a \star b$.

   (c) Perform and save in register 2 the result of c/d.

   (d) Add data in register 1 to data in register 2.

   (e) Move back the end result of step 4 to memory location r.

   Thus, a gain in execution time would result from the reduction in the data transfer time.

2. **Arithmetic operations within the processing unit:** In the case of (1.1), there are one multiplication, one division and one addition. Although each of these operations is of different nature and may require different execution times, we lump these together in a total of 3 "flops" (floating-point operations). Modern computer architectures are built to speed up the execution of these arithmetic operations by performing on

the same processor simultaneous computations. In such case, the last 5 steps implementation of (1.1) in which steps 2 and 3 are independent would become:

(a) Move the contents of a, b, c and d from memory to arithmetic unit.

(b) Perform and save in register 1 the result of $a \star b$; perform and save in register 2 the result of c/d.

(c) Add contents of register 1 to contents of register 2.

(d) Move back the end result of step 4 to memory location r.

Thus, the initial six steps necessary to execute (1.1) can be compacted into four steps.

It is well known that **linear algebra** concepts and operations are vector oriented. As we will see in future chapters, there are some linear algebra operations that occur frequently; they are the bricks on which the programs are built. These have been grouped under the term **Basic Linear Algebra Subprograms or BLAS** [24, 25, 44]. Since BLAS operations like matrix-matrix multiplication are so common, computer manufacturers have standardized them in libraries optimized for their machines. By incorporating these callable Fortran or C BLAS libraries in calculations, one can maintain a level of portability while still achieving significant fractions of optimal machine performance. Thus BLAS operations are made available to users, to whom their implementation is transparent and dependent on the specific compiler (or interpreter), which itself is tied up to the specific computer architecture. The standard classification of BLAS operations is done according to their level of complexity:

1. Level-1 BLAS: Vector-Vector operations,

2. Level-2 BLAS: Matrix-Vector operations,

3. Level-3 BLAS: Matrix-Matrix operations.

1.2 Matrix Notations

The fact that matrix computations are predominant in scientific computing, has led to inventing new software environments such as MATLAB [71] or Scilab, [60] which is a similar software but is a free and open source. In Table 1.1, are reported two ways for expressing the basic data in Linear Algebra. In this book, the two types are used depending on the context. However, sometimes MATLAB notations are used in mathematical expressions when it enhances understanding of the underlying mathematical concepts.

Mathematical Notation	MATLAB Notation
Column vector $x \in \mathbb{R}^n$	x
$|x| \in \mathbb{R}^n$, absolute value of a column vector	abs(x)
Row vector $r \in \mathbb{R}^{1 \times n} = x^T$	r=x'
$|r| \in \mathbb{R}^{1 \times n} = |x|^T$, absolute value of a row vector	r=abs(x)'
Matrix A (m by n)	A
$A_{ij} = A(i,j)$, coefficient of Matrix A	A(i,j)
$A^T \in \mathbb{R}^{n \times m}$, transpose matrix of $A \in \mathbb{R}^{m \times n}$	A'
$A_{ij}^T = A^T(i,j)$, coefficient of Matrix A^T	A'(i,j)
$A^* \in \mathbb{C}^{n \times m}$, adjoint (transpose of complex conjugate) of $A \in \mathbb{C}^{m \times n}$	A'
$A_{ij}^* = A^*(i,j)$, coefficient of Matrix A^*	A'(i,j)
$|A| \in \mathbb{R}_+^{m \times n}$ matrix of absolute values	abs(A)
$|A|_{ij} = |A(i,j)|$, coefficient of Matrix $|A|$	abs(A)(i,j)
$A_{.j}$, j^{th} column vector of A (m by n)	A(:,j)
$A_{i.}$, i^{th} row vector of A	A(i,:)
Sequence of row indices: $I = [i_1 i_2 ... i_k]$	I(1:k)
Sequence of column indices: $J = [j_1 j_2 ... j_l]$	J(1:l)

TABLE 1.1: Mathematical notations versus MATLAB notations.

These expressions can be combined in more complicated expressions. For instance, if I,J are permutation vectors on 1:k and 1:l, then the MATLAB instruction B=A(I,J) provides a matrix B which is obtained from the permutation of the rows and columns of A according to I and J. Mathematically, $B \in \mathbb{R}^{k \times l}$ with

$$B(r,s) = A(I(r), J(s)).$$

Remark 1.1 *Given an m by n matrix A and based on Table 1.1, we use for its column vectors either $\{A_{.j}| 1 \le j \le n\}$ or $\{A(:,j)|j = 1:n\}$. Similarly for the row vectors of A, we use either $\{A_{i.}| 1 \le i \le m\}$ or $A\{(i,:)|i = 1:m\}$.*

1.3   IEEE Floating Point Systems and Computer Arithmetic

In modern computer architectures, numbers are usually stored using a specific **IEEE**[1] notation, referred to as floating point notation (see [36]). In this book and as adopted by default in MATLAB, we consider the *IEEE double precision* notation, which provides the set $\mathbb{F}_d \subset \mathbb{R}$ of "double precision

[1]IEEE: Institute of Electrical and Electronics Engineers.

computer numbers," whereas if $x \in \mathbb{F}_d$, then:

$$\text{if } x \neq 0,\ x = \pm(1.f)_2 \times 2^{c-1023},\ 1 \leq c \leq 2046$$

with, by convention:

$$\pm 0 \text{ represented by } c = 0 \text{ and } f = 0.$$

The system $\mathbb{F}_d$ uses a word of 8 bytes with f and c represented, respectively, by 52 and 11 bits with one final bit to the sign of the number x. Thus if x_1 and x_2 are adjacent numbers in $\mathbb{F}_d$ then:

$$\frac{|x_1 - x_2|}{\max\{|x_1|, |x_2|\}} \leq \epsilon_M (\text{ the epsilon machine}),$$

with $\epsilon_M = 2^{-52} \approx 2.2 \times 10^{-16}$. Consequently, each exact arithmetic operation (in infinite precision arithmetic):

$$\cdot \in \{+, -, \star, /\},$$

is simulated by a built-in computer arithmetic operation:

$$\odot \in \{\oplus, \ominus, \circledast, \oslash\},$$

that is subject to a roundoff error. Specifically:

$$\forall x, y \in \mathbb{F}_d : x \cdot y \in \mathbb{R} \text{ and } x \odot y \in \mathbb{F}_d,$$

with:

$$\forall x, y \in \mathbb{F}_d : |x \cdot y - x \odot y| = O(\epsilon_M) \Longleftrightarrow x \odot y = (1+\gamma)x \cdot y,\ |\gamma| = O(\epsilon_M). \quad (1.2)$$

For more details on floating-point systems, we refer the reader to [36] and Chapter 1 of [28].

1.4   Vector-Vector Operations: Level-1 BLAS

To increase the speed of executing arithmetic operations, it is suitable to use, in modern computer architecture, transfer of data between memory and processor in "blocks" of numbers. This is referred to as "vectorized" data processing and is implemented on vector and parallel computers.

These computers are capable of implementing in an efficient manner a concise set of vector operations. Let x, y and $z \in \mathbb{R}^n$ and $a \in \mathbb{R}$ (although in practice x, y and $z \in \mathbb{F}_d^n$ and $a \in \mathbb{F}_d$.) The most common linear algebra vector-vector operations are displayed in Table 1.2.

Operation	Vector-Vector Notation	Component Notation	MATLAB Implementation
Scalar-Vector Multiplication	$z = ax$	$z_i = ax_i$	z=a*x
Vector Addition	$z = x + y$	$z_i = x_i + y_i$	z=x+y
saxpy	$z = ax + y$	$z_i = ax_i + y_i$	z=a*x+y
scalar product	$a = x^T y$	$a = \sum_i x_i y_i$	z=x'*y
Hadamard vector multiply (component-wise)	$z = x \ .\times\ y$	$z_i = x_i y_i$	z=x.*y

TABLE 1.2: Basic vector-vector operations.

In this table, `saxpy` means "scalar $a \star x$ plus y"; it is not independent of the preceding two operations, but it is common to list it as a basic operation because it occurs frequently in practical calculations.

The analysis of complexity would go as follows. By denoting n the length of the involved vectors, the number n_f of floating point operations – or flops – are reported in Table 1.3. In the same table are also indicated the corresponding number n_r of memory references, and the ratio $q = \frac{n_f}{n_r}$.

Operation	n_f: **Number of Flops**	n_r: **Number of References**	$q = n_f/n_r$
Scalar-Vector Multiplication	n	$2n + 1$	$\approx 1/2$
Vector Addition	n	$3n$	$\approx 1/3$
saxpy	$2n$	$3n + 1$	$\approx 2/3$
scalar product	$2n$	$2n + 1$	≈ 1
vector multiply (component-wise)	n	$3n$	$\approx 1/3$

TABLE 1.3: Complexities in flops and memory references (Level-1 BLAS).

It is instructive to classify the BLAS according to the number of floating point operations and the number of memory references required for a basic linear algebra operation. The parameter q is the ratio of flops to memory references. Generally, when $q > 1$, a BLAS operation would require more useful processor work than time spent moving data, while when $q < 1$, it would be otherwise, more transfer of data than processor work. If $n_f = O(n^\alpha)$ and $n_r = O(n^\beta)$, BLAS operations are classified according to the "level" of α.

In this case of vector operations, the BLAS operations in Table 1.3 are considered to be of level 1.

Remark 1.2 Effect of Rounding Errors in Level-1 BLAS

Floating-point arithmetic in implementing Level-1 BLAS operations leads to rounding errors. Specifically:

• In *saxpy* if z_c is the computed result of $z = ax + y$, then using (1.2):

$$\forall i, \; z_{c,i} \; = \; fl(fl(a \times x_i) + y_i) = (fl(a \times x_i) + y_i)(1 + \epsilon_1)$$
$$= \; ((a \times x_i)(1 + \epsilon_2) + y_i)(1 + \epsilon_1).$$

Hence:

$$z_{c,i} = (a \times x_i + y_i)(1 + \epsilon_1) + ax_i\epsilon_1(1 + \epsilon_2) = z_i(1 + \epsilon_1) + ax_i\epsilon_2(1 + \epsilon_1),$$

and therefore:

$$z_{c,i} - z_i = \epsilon_1 z_i + ax_i\epsilon_2(1 + \epsilon_1).$$

This implies:

$$|z_{c,i} - z_i| \le \epsilon_M |z_i| + |ax_i|\epsilon_M(1 + \epsilon_M),$$

i.e.,

$$|z_{c,i} - z_i| \le \epsilon_M(|a||x_i| + |y_i|) + (|a||x_i| + |y_i|)\epsilon_M(1 + \epsilon_M),$$

Hence, one concludes with the following proposition:

Proposition 1.1

$$|z_{c,i} - z_i| \le 2(|a||x_i| + |y_i|)O(\epsilon_M), \; \forall i = 1, ..., n, \tag{1.3}$$

i.e.,

$$|z - z_c| \le 2(|a||x| + |y|)O(\epsilon_M)$$

∎

• For a finite sum of numbers, $s = \sum_{i=1}^{n} x_i$ and its corresponding floating-point operation, $s_c = fl(\sum_{i=1}^{n} x_i)$, we show:

Proposition 1.2

$$\left| fl(\sum_{i=1}^{n} x_i) - \sum_{i=1}^{n} x_i \right| \le (n \sum_{i=1}^{n} |x_i|)O(\epsilon_M). \tag{1.4}$$

Proof. Starting with $fl(x_1 + x_2) = (x_1 + x_2)(1 + \epsilon_1)$, $|\epsilon_1| \le \epsilon_M$, then:

$$fl(x_1 + x_2 + x_3) = fl(fl(x_1 + x_2) + x_3) = (fl(x_1 + x_2) + x_3)(1 + \epsilon_2),$$

i.e.,

$$fl(x_1 + x_2 + x_3) = ((x_1 + x_2)(1 + \epsilon_1) + x_3)(1 + \epsilon_2).$$

Hence:

$$fl(x_1 + x_2 + x_3) = (x_1 + x_2 + x_3) + (x_1 + x_2)(\epsilon_1 + \epsilon_2 + \epsilon_1\epsilon_2) + x_3\epsilon_2.$$

Therefore:

$$|fl(x_1 + x_2 + x_3) - (x_1 + x_2 + x_3)| \le |x_1 + x_2|(2\epsilon_M + \epsilon_M^2) + |x_3|\epsilon_M,$$

and:

$$|fl(x_1 + x_2 + x_3) - (x_1 + x_2 + x_3)| \le (|x_1| + |x_2| + |x_3|)(3\epsilon_M + \epsilon_M^2).$$

The estimate (1.4) is reached by induction. ∎

• Consider now the scalar product $a = x^T y = \sum_{i=1}^{n} x_i y_i$ and its associated floating point operation $a_c = fl(x^T y)$. Letting:

$$z = \{z_i | \ i = 1, ..., n, \}, \ z_i = fl(x_i y_i)$$

one has $a_c = fl(\sum_{i=1}^{n} z_i)$. Using (1.4) in Proposition 1.2, one obtains:

$$\left| fl\left(\sum_{i=1}^{n} z_i\right) - \sum_{i=1}^{n} z_i \right| \le \left(n \sum_{i=1}^{n} |z_i|\right) O(\epsilon_M).$$

Given that:

$$fl(x^T y) - x^T y = fl\left(\sum_{i=1}^{n} z_i\right) - \sum_{i=1}^{n} z_i + \sum_{i=1}^{n} (fl(x_i y_i) - x_i y_i)$$

and as $fl(x_i y_i) - x_i y_i = x_i y_i \epsilon_M$, one obtains:

$$|fl(x^T y) - x^T y| \le \left(n \sum_{i=1}^{n} |z_i|\right) O(\epsilon_M) + \sum_{i=1}^{n} |x_i||y_i|\epsilon_M.$$

Moreover:

$$\sum_{i=1}^{n} |z_i| \le \sum_{i=1}^{n} |x_i||y_i|(1 + \epsilon_M).$$

Hence by letting $|x|^T |y| = \sum_{i=1}^{n} |x_i||y_i|$, one obtains:

$$|a_c - a| = |fl(x^T y) - x^T y| \le |x|^T |y| O(n\epsilon_M). \tag{1.5}$$

The completion of this inequality is left to Exercise 1.12. ∎

1.5   Matrix-Vector Operations: Level-2 BLAS

Let $x \in \mathbb{R}^n$, $y \in \mathbb{R}^m$, $z \in \mathbb{R}^m$ and $A \in \mathbb{R}^{m \times n}$. The two most common linear algebra matrix-vector operations are expressed in Table 1.4.

Operation	Vector Notation	Scalar Notation	MATLAB
Matrix-Vector Multiplication (GAXPY)	$z = Ax + y$	$z_i = \sum_{j=1}^{n} A_{ij}x_j + y_i$	z=A*x+y
external product	$A = xy^T$	$A_{ij} = x_i y_j$	A=x*y'

TABLE 1.4: Basic matrix-vector operations.

In the table, `GAXPY` means "General <u>AX</u> <u>P</u>lus <u>Y</u>". Given the formula:

$$z(i) = \sum_{j=1}^{n} A(i,j)x(j) + y(i),\ i = 1, ..., m. \qquad (1.6)$$

(1.6) can be rewritten as:

$$z = z(1:m) = \sum_{j=1}^{n} A(1:m,j)x(j) + y(1:m) = \sum_{j=1}^{n} x(j) \times A(:,j) + y.$$

In case the matrix A is stored column-wise, the implementation of such operation uses saxpy's as in Algorithm 1.1.

Algorithm 1.1 GAXPY Implementation Using Saxpy

```
function z=GAXPY1(A,x,y)
[m,n]=size(A);n1=length(x);m1=length(y);
if m1==m & n1==n
    x=x(:);y=y(:);z=zeros(m,1);
    for j=1:n
        z=x(j)*A(:,j)+z;
    end
    z=z+y;
end
```

Similarly, (1.6) can be rewritten by using scalar products, particularly when the matrix A is stored row-wise, specifically:

$$z(i) = A(i,:) * x + y(i),\ i = 1:m.$$

In such case, one gets Algorithm 1.2.

Algorithm 1.2 GAXPY Implementation Using Scalar Product

```
function z=GAXPY2(A,x,y)
[m,n]=size(A);n1=length(x);m1=length(y);
if m1==m & n1==n
   x=x(:);y=y(:);z=zeros(m,1);
   for i=1:m
     z(i)=A(i,:)*x;
   end
   z=z+y;
end
```

Both implementations lead to the same number of arithmetic operations:

$$\sum_{i=1}^{m} (2n+1) = 2nm + m.$$

For external products, note that if $A = xy^T$, then this can be translated in a scalar form as:

$$A_{ij} = x_i y_j, \; i = 1, \cdots, m, \; j = 1, \cdots, n,$$

or in vector form (column-wise) as:

$$\mathtt{A}(:,\mathtt{j}) = \mathtt{y}(\mathtt{j}) * \mathtt{x}, \; \mathtt{j} = 1 : \mathtt{n},$$

or in vector form (row-wise):

$$\mathtt{A}(\mathtt{i},:) = \mathtt{x}(\mathtt{i}) * \mathtt{y}', \; \mathtt{i} = 1 : \mathtt{m}.$$

In the case of square matrices, i.e., $n = m$, one gets Table 1.5.

Operation	n_f: **Number of Flops**	n_r: **Number of References**	$q = n_f/n_r$
GAXPY 1 and 2	$2n^2 + n$	$n^2 + 3n$	≈ 2
external product	n^2	$2n + n^2$	≈ 1

TABLE 1.5: Complexities in flops end memory references (Level-2 BLAS).

Remark 1.3 Effect of Rounding Errors in Level-2 BLAS

In such case one has:

• In *GAXPY* if z_c is the computed result of $z = Ax + y$ where A is an n by n matrix, then given that each component of z is computed using a scalar product, one has:

$$z_i = A_{i.} x + y_i.$$

Hence:

$$z_{c,i} = fl(fl(A_{i.}x) + y_i) = fl(A_{i.}x) + y_i + \delta_i = A_{i.}x + y_i + \gamma_i + \delta_i = z_i + \gamma_i + \delta_i.$$

Using (1.3) and (1.5), one obtains:

$$|z - zc| \leq n(|A||x| + |y|)O(\epsilon_M) \qquad (1.7)$$

The proof of this inequality is left to Exercise 1.13. ∎

• For vector external product, if A_c is the computed result of $A = xy^T$, then using also (1.2), one easily obtains:

$$(A_c)(i, j) = A(i, j)(1 + \gamma_{ij}), \ |\gamma_{ij}| = O(\epsilon_M).$$

1.6 Matrix-Matrix Operations: Level-3 BLAS

Note that matrix additions and multiplication by a scalar, typically of the form: $aA + B$, $a \in \mathbb{R}$, $A, B \in \mathbb{R}^{n \times n}$ are level-2 BLAS, since they require $O(n^2)$ arithmetic operations.

On the other hand, matrix multiplications are level-3 BLAS. Let $A \in \mathbb{R}^{m \times n}$, $B \in \mathbb{R}^{n \times p}$. The product of the two matrices $C = A \times B$ is such that $C \in \mathbb{R}^{m \times p}$, with:

$$C_{ij} = \sum_{k=1}^{n} A_{ik} B_{kj}, \ 1 \leq i \leq m, \ 1 \leq j \leq p. \qquad (1.8)$$

Such operation would require the evaluation of mp coefficients, each term of these being the sum of n products. Hence the total number of operations is:

$$2mpn.$$

When $m = p = n$, this becomes $2n^3$. As for the number of references, it is obviously:

$$mn + np + mp.$$

When $m = p = n$, the number of references becomes $3n^2$. In such case, one gets Table 1.6.

Operation	n_f: **Number of Flops**	n_r: **Number of References**	$q = n_f/n_r$
Matrix Multiplications	$2n^3$	$3n^2$	$= 2n/3$

TABLE 1.6: Complexities in flops end memory references (Level-3 BLAS).

As one can check in such case, there are more processor requirements than data transfers. This is a happy situation where the implementation can be organized in a way where the memory bandwidth does not slow down the computation: a ratio $q = O(n)$ indicates that it should be possible to re-use data and therefore to store them in a local memory with fast access (a cache memory); this is the situation when the three matrices can fit in the cache memory.

1.6.1   Matrix Multiplication Using GAXPYs

The matrix multiplication $C = AB$ is equivalent to:

$$C = [C(:,1)\, C(:,2)...\, C(:,p)] = A \times [B(:,1)\, B(:,2)\, ...\, B(:,p)]$$

resulting in:

$$C(:,j) = A \times B(:,j) = \sum_{k=1}^{n} B(k,j)A(:,k),\ j = 1,...,p \qquad (1.9)$$

If $A \in \mathbb{R}^{m \times n}$ and $B \in \mathbb{R}^{n \times p}$ are stored column-wise and the product matrix $C = A \times B$ is also stored column-wise. In such case, the formula (1.8) can be implemented using GAXPY's operations (1.21) that in turn are based on SAXPY's operations. These are suited when both matrices A and B are stored column-wise.

Algorithm 1.3 Matrix Product Implementation Using GAXPY

```
function C=MatMult1(A,B)
[m,n]=size(A);[n1,p]=size(B);
if n==n1
    for j=1:p
        C(:,j)=A*B(:,j);
    end
end
```

Note that the segment of program:

```
for j=1:p
    C(:,j)=A*B(:,j);
end
```

is implemented using saxpy's:

```
C=zeros(m,p);
for j=1:p
    for k=1:n
        C(:,j)=B(k,j)*A(:,k)+C(:,j);
    end
end
```

1.6.2   Matrix Multiplication Using Scalar Products

We now interpret formula (1.8) as follows:

$$C_{ij} = A(i, :)B(:, j) \tag{1.10}$$

which is the scalar product of the ith row of A with the jth column of B. For implementation, check Exercises 1.1 and 1.2 at the end of the chapter.

1.6.3   Matrix Multiplication Using External Products

We begin by rewriting (1.8):

$$C = C(:, :) = \sum_{k=1}^{n} A(:, k)B(k, :). \tag{1.11}$$

Note that $A(:, k)B(k, :)$ is an **external product**, since $A(:, k) \in \mathbb{R}^{m,1}$ and $B(k, :) \in \mathbb{R}^{1,p}$. On that basis, do Exercises 1.3 and 1.4 at the end of the chapter.

1.6.4   Block Multiplications

Consider the matrices $A \in \mathbb{R}^{m \times n}$ and $B \in \mathbb{R}^{n \times p}$. Let $m_1 < m$, $n_1 < n$ and $p_1 < p$, with $m_2 = m - m_1$, $n_2 = n - n_1$ and $p_2 = p - p_1$. Consider the following subdivision of A and B:

$$A = \left(\begin{array}{c|c} A_{11} & A_{12} \\ \hline A_{21} & A_{22} \end{array} \right) \tag{1.12}$$

$$B = \left(\begin{array}{c|c} B_{11} & B_{12} \\ \hline B_{21} & B_{22} \end{array} \right), \tag{1.13}$$

where $A_{11} \in \mathbb{R}^{m_1 \times n_1}$, $A_{12} \in \mathbb{R}^{m_1 \times n_2}$, $A_{21} \in \mathbb{R}^{m_2 \times n_1}$, $A_{22} \in \mathbb{R}^{m_2 \times n_2}$ and $B_{11} \in \mathbb{R}^{n_1 \times p_1}$, $B_{12} \in \mathbb{R}^{n_1 \times p_2}$, $B_{21} \in \mathbb{R}^{n_2 \times p_1}$, $B_{22} \in \mathbb{R}^{n_2 \times p_2}$. Then the matrix product $C = A \times B$ can be written as:

$$C = \left(\begin{array}{c|c} C_{11} & C_{12} \\ \hline C_{21} & C_{22} \end{array} \right), \tag{1.14}$$

with $C_{11} \in \mathbb{R}^{m_1 \times p_1}$, $C_{12} \in \mathbb{R}^{m_1 \times p_2}$, $C_{21} \in \mathbb{R}^{m_2 \times p_1}$, $C_{22} \in \mathbb{R}^{m_2 \times p_2}$. Furthermore, one has the following relations:

$$
\begin{array}{rll}
C_{11} & = & A_{11}B_{11} + A_{12}B_{21}, & \quad (1.15) \\
C_{12} & = & A_{11}B_{12} + A_{12}B_{22}, & \quad (1.16) \\
C_{21} & = & A_{21}B_{11} + A_{22}B_{21}, & \quad (1.17) \\
C_{22} & = & A_{21}B_{12} + A_{22}B_{22}. & \quad (1.18)
\end{array}
$$

Breaking the multiplication by blocks allows one to distribute it through a computer architecture that has multi-processors. For example the preceding formulae can be implemented using two processors with the first computing C_{11}, C_{21} and the second computing C_{12}, C_{22} or even four processors with each one building one of the four blocks.

1.6.5   An Efficient Data Management

Let us consider an implementation strategy for the following dense matrix multiplication primitive,

$$C \; := \; C \; + \; A \; * \; B \qquad\qquad (I)$$

on a one-processor equipped with cache memory. This type of memory is characterized by its fast access, i.e., reading one word costs only one clock period. Since the storage capacity of a cache memory is limited, the goal of a code developer is to re-use, as much as possible, data stored in the cache memory. The following analysis is an adaptation of the discussion in [31] for a uniprocessor.

Let M be the storage capacity of the cache memory and let us assume that matrices A, B and C are, respectively, $n_1 \times n_2$, $n_2 \times n_3$ and $n_1 \times n_3$ matrices. Partitioning these matrices into blocks of sizes $m_1 \times m_2$, $m_2 \times m_3$ and $m_1 \times m_3$, respectively, where $n_i = m_i * k_i$ for all $i = 1,\ 2,\ 3$, our goal is then to estimate the block sizes m_i which maximize data re-use under the cache size constraint. Instruction (I) can be expressed as the nested loop,

```
for i = 1: k1
  for k = 1: k2
    for j = 1: k3
      C(i,j) = C(i,j) + A(i,k) * B(k,j);
    end
  end
end
```

where `Cij`, `Aik` and `Bkj` are, respectively, blocks of C, A and B, obviously indexed. The innermost loop j, contains the block A_{ik} which has to be contained in the cache, i.e.,

$$m_1 m_2 \leq M.$$

Further, since the blocks are obviously smaller than the original matrices, the additional constraints are: $1 \leq m_i \leq n_i$ for $i = 1,\ 2,\ 3$.

Evaluating the volume of the data migrations using the number of data loads necessary for the whole procedure (assuming that the above constraints are satisfied) we have: (i) all the blocks of the matrix A are loaded only once; (ii) all blocks of the matrix B are loaded k_1 times, and (iii) all the blocks of the matrix C are loaded k_2 times. Thus the total amount of loads is given by:

$$L = n_1 n_2 + n_1 n_2 n_3 \left(\frac{1}{m_1} + \frac{1}{m_2} \right)$$

Choosing m_3 freely, the values of m_1 and m_2, which minimize $\frac{1}{m_1} + \frac{1}{m_2}$ under the previous constraints, are obtained as follows:

```
if n2*n1<=M
      m1=n1;m2=n2;
elseif n2<=M
      m1=M/n2;m2=n2;
elseif n1<=sqrt(M)
      m1=n1;m2=M/n1;
else
      m1=sqrt(M);m2=sqrt(M);
end
```

In practice, however, M should be slightly smaller than the total cache volume to allow for storing few additional vectors.

Remark 1.4 Effect of Rounding Errors in Matrix Multiplication

If one implements $C = A \times B$ on square n by n matrices using scalar products with the computed matrix result being C_c, one then has $C_{ij} = A_{i.} B_{.j}$ and $C_{c,ij} = fl(A_{i.} B_{.j})$. Using (1.5), one has:

$$|C_{ij} - C_{c,ij}| \le n |A|_{i.} |B|_{.j} O(\epsilon_M),$$

i.e.,

$$|C - C_c| \le n |A|.|B| O(\epsilon_M), \tag{1.19}$$

1.7 Sparse Matrices: Storage and Associated Operations

In many numerical simulations, a matrix of huge order is obtained in several applications, such as when discretizing a Partial Differential Equation. Typically, the order n is often higher than one million. The usual storage of a square matrix as a 2-dimensional array would require n^2 memory words (so-called dense storage). This is impossible on usual computers. In usual situations, most of the entries are zero in such matrices and it becomes tractable to only store the nonzero entries with some information on their position in the matrix. This type of storage is called sparse.

To illustrate such situations, let us look respectively at the 3-point and 5-point discretization of the Laplacian operator in 1 and 2 dimensions.

In the first case, with p grid points one gets a tridiagonal matrix of order $n = p$ with $n_z = 3n-2$ non-zero entries. On the other hand, in a 2-dimensional square grid of $p \times p$ grid points and by numbering the nodes by rows, one gets a matrix of order $n = p^2$, in which every row has a maximum of 5 non-zero entries. The non-zero entries are then located on 5 non-zero diagonals that are not contiguous as in the tridiagonal case. Independently, the number n_z of non-zeros entries is $n_z = 5\, n + O(1)$. It can be noticed that, since the density of the non-zero entries is equal to $\tau = \frac{5}{n} + O(\frac{1}{n^2})$, the density decreases when refining the mesh and therefore increasing p. It then becomes clear that a sparse storage is compulsory.

The pattern of a sparse matrix[2] is the collection of all the locations of the non-zeros entries. The pattern of A is usually represented by a graph $\mathcal{G}(K, \mathcal{E})$ where $K = \{1, \cdots, n\}$ is the set of vertices and $\mathcal{E} = \{(i, j) \in K \times K \mid A_{ij} \neq 0\}$ the set of edges.

Renumbering the rows and the columns is possible by considering permutations.

Common storages are as follows:

Band storage: An m by n band matrix A with l lower diagonals and u upper diagonals may be stored compactly in a two-dimensional array Ab with $l + u + 1$ rows and n columns. Columns of the matrix A are stored in corresponding columns of the array Ab, and diagonals of the matrix A are stored in the rows of the array Ab. More precisely, $A(i, j)$ is stored in $Ab(u+1+i-j, j)$ for

$$\max\{1, j - ku\} \leq i \leq \min\{m, j + kl\}.$$

Such storage scheme is recommended when using LAPACK routines, particularly, whenever $\max\{l, u\} << \min\{m, n\}$. For example, [5], when $m = n = 5$, $l = 2$ and $u = 1$, the transformation from A_b to Ab is displayed in Figure 1.1.

5 by 5 banded matrix A	Storage of A in 4 by 5 array
$\begin{pmatrix} a_{11} & a_{12} & 0 & 0 & 0 \\ a_{21} & a_{22} & a_{23} & 0 & 0 \\ a_{31} & a_{32} & a_{33} & a_{34} & 0 \\ 0 & a_{42} & a_{43} & a_{44} & a_{45} \\ 0 & 0 & a_{53} & a_{54} & a_{55} \end{pmatrix}$	$\begin{pmatrix} \text{unused} & a_{12} & a_{23} & a_{34} & a_{45} \\ a_{11} & a_{22} & a_{33} & a_{44} & a_{55} \\ a_{21} & a_{32} & a_{43} & a_{54} & \text{unused} \\ a_{31} & a_{42} & a_{53} & \text{unused} & \text{unused} \end{pmatrix}$

FIGURE 1.1: Typical two-dimensional storage of a banded matrix

---

[2]The word "sparse" is theoretically referring to a compressed storage, but very often "sparse matrix" is used to mention that the matrix must be sparsely stored because most of its entries are zeros.

Thus, the storage is organized by diagonals: The rectangular array Ab stores row-wise all the diagonals included in the band of the original matrix A which is limited by the two extreme diagonals that include non-zero entries. The bandwidth b is given by the number of rows of Ab. For the tridiagonal matrix case, the bandwidth is $b = 3$, while for 5-point discretization, the bandwidth is $b = 2p + 1$ and the necessary storage amounts to $(2p+1)n = O(n^{3/2})$. This storage is very regular but it stores many zeros (many of the stored diagonals of the band are zero-diagonals!).

Obviously, when the pattern of the matrix is not structured by diagonals as in the previous two examples, loss of storage efficiency in storing unnecessary zeros can become considerably more important. For that reason, one looks for alternatives that allow "compressed" storages that exclusively store non-zero coefficients.

Compressed storage by coordinates (COO): This type of storage is used in `MATLAB` and is based on the principle of storing all non-zero entries in a 1-dimensional array V of length n_z. Correspondingly, row indices and column indices of these non-zero coefficients are stored in the same order in two other 1-dimensional arrays I and J of the same length n_z. The location of the non-zeros entries is characterized by:
`A(I(k),J(k))=V(k), k=1: `n_z.

Compressed storage by columns (CSC): In this type of storage, we modify the COO way by keeping both the two vector arrays of length n_z, V and I that store respectively the non-zero coefficients of the matrix and the associated row indices. The COO vector J of length n_z is replaced by one of length $n + 1$ which role is to "point" at the row vector I in the following way:

- For $j = 1, 2, ...n$, $i_j = J(j)$ is the position in the vector I of first row index in the jth column of A such that $A(I(i_j), j)$ is stored $(A(I(i_j), j) \neq 0)$.

- For the sake of consistency, we let $J(1) = 1$ and $J(n + 1) = n_z + 1$.

- Accordingly:

$$\text{For } j = 1, ..., n : 1 \leq i_j < i_{j+1} \leq n_z + 1,$$

and

$$A(I(k), j) = V(k) \neq 0, \ i_j \leq k \leq i_{j+1} - 1.$$

Hence, in `MATLAB`, for any column $j = 1 : n$, the non-zeros entries are given by: `A(I(k),j)=V(k), for k=J(j):J(j+1)-1`.

Compressed storage by rows (CSR): This storage is similar to the previous one but by considering the rows instead of the columns. It can be noticed that a CSC storage of the matrix A is the CSR storage of the transpose matrix A^T. The details of the implementation are left to Exercise 1.19.

In what follows, we will adopt `MATLAB` sparse storage and we rewrite BLAS functions according to this type of storage by considering BLAS-2 and BLAS-3 procedures that involve matrix-vector and matrix-matrix multiplications.

Sparse GAXPY

As an illustration, given A a sparse matrix n by n and two vectors x, $y \in \mathbb{R}^n$ we write the MATLAB routine that gives $z = Ax + y$. The loop involves:

1. An extraction *gather* (indirect loads) of the 3 columns data (COO) describing the sparse matrix A.

2. A *scatter* of data (indirect saves) in z(I(i))=z(I(i))+a(i)*x(J(i)), to assemble the resulting vector z.

Note that the scatter-gathers are optimized on modern computer architectures.

The MATLAB version of such operation is given in Algorithm 1.4. If $n_z = n_z(A)$ is the maximum number of non-zero coefficients in A, then the number of flops n_f to execute SGAXPY is given by $n_f = 2n_z$, which is only $O(n_z)$.

Algorithm 1.4 Sparse GAXPY Implementation Using Scalar Products

```
function z=SGAXPY(A,x,y)
% Input: Sparse n by n matrix A
%        Two column vectors x and y of length n
% Output: z=A*x+y
z=y;
[I,J,a]=find(A);
% In MATLAB, given a sparse matrix A
% [r,c,v] = find(A) returns
% r and c: row and column indices vectors,
% and correspondingly a vector
% v of the associated nonzero values, i.e.,
% A(r(I),c(I))=v(I)
k=length(a);
for i=1:k
    z(I(i))=z(I(i))+a(i)*x(J(i));
end
```

Sparse Matrix Multiplication

A second illustration is given when considering the matrix multiplication $C = A \times B$ with both A and B sparse n by n matrices. It is left to be worked out in Exercise 1.18.

1.8   Exercises

Exercise 1.1

Write the MATLAB program $MatMult2$ that implements the multiplication $C = AB$ using (1.10). Discuss if the data transfer using this approach is efficient.

Exercise 1.2

Transform $MatMult2$ to implement the product $C = A^T B$, where $A \in \mathbb{R}^{m \times n}$ and $B \in \mathbb{R}^{m \times p}$. What do you conclude in terms of data transfer efficiency?

Exercise 1.3

Write the MATLAB program $MatMult3$ that implements the multiplication $C = AB$ using external products as in (1.11). Discuss the data transfer efficiency.

Exercise 1.4

Transform $MatMult3$ to implement the multiplication $C = AB^T$, where $A \in \mathbb{R}^{m \times n}$ and $B \in \mathbb{R}^{p \times n}$. What do you conclude in terms of data transfer efficiency?

In the next exercises, we consider square $n \times n$ matrices; I is the identity matrix (In MATLAB eye(n) gives a square n by n matrix). If e_k is the column unit vector which components are all 0's except the kth component which is equal to 1, i.e.,

$$e_1 = [1\,0\,0...0]^T, e_1 = [0\,1\,0...0]^T, ..., e_1 = [0\,0\,0...1]^T,$$

then the identity matrix I is such that:

$$I = \begin{pmatrix} 1 & 0 & 0 & 0 & ... & 0 \\ 0 & 1 & . & 0 & ... & 0 \\ . & ... & . & . & ... & . \\ 0 & ... & 0 & 0 & 1 & 0 \\ 0 & ... & ... & 0 & 0 & 1 \end{pmatrix} = (e_1\, e_2\, ...e_n) = \begin{pmatrix} e_1^T \\ e_2^T \\ ... \\ e_n^T \end{pmatrix}$$

To generate in MATLAB the unit (column) vector $e_i \in \mathbb{R}^n$, one may uses the sequence of commands:
I=eye(n);e(i)=I(:,i).

Exercise 1.5

Let $A \in \mathbb{R}^{n \times n}$, $x \in \mathbb{R}^k$. Find the first column of

$$M = (A - x_k I)(A - x_{k-1}I)...(A - x_1 I),$$

using a sequence of GAXPY's operations.

Exercise 1.6

Give the algorithm that computes $(xy^T)^k$ where x and y are n vectors.

Exercise 1.7

Let A be a square n by n matrix. Write the algorithms that:

1. Double column 1
2. Multiply by c column j, $j = 1...n$.
3. Multiply **simultaneously** columns 1,2,...,n by $c_1, c_2, ..., c_n$.
4. Multiply **simultaneously** rows 1,2,...,n by $c_1, c_2, ..., c_n$.
5. Multiply row 1 by a constant c and add to row 2, placing the result in row 2.
6. Interchange rows 1 and 2 and more generally rows i and j.

Exercise 1.8

The `MATLAB` `diag` command allows to generate square diagonal matrices from a column (or row vector) $c \in \mathbb{R}^n$. `D=diag(c)` gives a square matrix:

$$D = \begin{pmatrix} c_1 & 0 & 0 & 0 & ... & 0 \\ 0 & c_2 & . & 0 & ... & 0 \\ . & ... & . & . & ... & . \\ 0 & ... & 0 & 0 & c_{n-1} & 0 \\ 0 & ... & ... & 0 & 0 & c_n \end{pmatrix}$$

Given a square matrix $A \in \mathbb{R}^{n \times n}$, prove that DA results in multiplying simultaneously each row i of A by $c(i)$ and that AD would have the same effect on the columns of A. Note that the same `diag` command allows to extract diagonal vectors from matrices.

Exercise 1.9

Let A and B be square matrices (both stored column-wise) in $\mathbb{R}^{n \times n}$ with B an **Upper Triangular** matrix. Consider the formula that gives $C = A \times B$:

$$C(:,j) = \Sigma_{k=1}^{n} B(k,j) \times A(:,k) \tag{1.20}$$

1. Use and modify (1.20) to obtain an algorithm that computes $C = A \times B$ and **exploits the special structure** of B (no multiplications by zeros). Give your answer by filling in the missing spaces in the following `MATLAB` program that gives the matrix C **column-wise**:

```
function C=MultRegUpper(A,B)
C=zeros(n,n);
for j=1:n
    v=B(  ,j);
```

```
    for k=
        C(  ,  )=C(  ,  )+v(  )*A(  ,  );
    end
end
```

2. Assume A is **also an upper triangular matrix**. Modify (1.20) to obtain an algorithm that computes $C = A \times B$ **exploiting the special structures** of A and B. Give your answer by filling in the missing spaces in the following MATLAB program that gives the matrix C **column-wise**:

```
function C=MultUpperUpper(A,B)
C=zeros(n,n);
for j=1:n
    v=B( ,j);
    for k=
        C(  ,  )=C(  ,  )+v(  )*A(  ,  );
    end
end
```

Give the number of flops needed to execute this algorithm.

Exercise 1.10

Let A and B be square matrices (both stored column-wise) in $\mathbb{R}^{n \times n}$ with B a **Lower Triangular** matrix. Consider the formula that gives $C = A \times B$:

$$C(:,j) = \Sigma_{k=1}^{n} B(k,j) \times A(:,k), \ j = 1, ..., n \qquad (1.21)$$

1. Use and modify (1.21) to obtain an algorithm that computes $C = A \times B$ and **exploits the special structure** of B (no multiplications by zeros). Give your answer by filling in the missing spaces in the following MATLAB program that gives the matrix C **column-wise**:

```
function C=MultRegLower(A,B)
C=zeros(n,n);
for j=1:n
    v=B( ,j);
    for k=
        C(  ,  )=C(  ,  )+v(  )*A(  ,  );
    end
end
```

2. Assume A is **also a lower triangular matrix**. Modify (1.20) to obtain an algorithm that computes $C = A \times B$ **exploiting the special structures** of A and B. Give your answer by filling in the missing spaces in the following MATLAB program that gives the matrix C **column-wise**:

```
function C=MultLowerLower(A,B)
C=zeros(n,n);
for j=1:n
```

```
        v=B(  ,j);
        for k=
            C(   ,   )=C(   ,   )+v(   )*A(   ,   );
        end
    end
```

Give the number of flops needed to execute this algorithm.

Exercise 1.11 *Multiplication of matrices using diagonals.*

For any matrix $A \in \mathbb{R}^{n \times n}$ and any integer d such that $-(n-1) \leq d \leq n-1$, the diagonal d of the matrix A is the vector D_d defined by:

- if $d = 0$: $D_0 = (A_{11}, \cdots, A_{n,n})$.

- if $d > 0$: $D_d = (A_{1,d+1}, \cdots, A_{n-d,n})$.

- if $d < 0$: $D_d = (A_{|d|+1,1}, \cdots, A_{n,n-|d|})$.

In MATLAB, the command `diag(A,d)` determines the diagonal d of A (it can also be called when A is a rectangular matrix). Conversely, the command `A=diag(a,d)`, where a is a vector of length k, creates a square matrix of order $k + |d|$ whose diagonal d is a and all other diagonals are zeros.

1. Prove that the position (i, j) belongs to the diagonal d if and only if $j - i = d$.

2. In MATLAB, define two matrices `A=diag(a,d)` and `B=diag(b,e)` where a and b are two vectors of respective length k and ℓ satisfying $n = k+d = \ell + e$. Observe that `C=A*B` is a matrix where the only possible non-zero diagonal is the diagonal $f = d + e$. Compute also `B*A`. Conclude.

3. Prove the previous result and determine the entries of the diagonal $f = d + e$ of the resulting matrix `C`.

4. In MATLAB, given `A=diag(a,d)` and `B=diag(b,e)` where a and b are two vectors of respective length k and ℓ satisfying $n = k + d = \ell + e$, create the function `[c,f]=mmultd(a,d,b,e,n)` such that `C=diag(c,f)` where `C=A*B`.

5. Let `A` and `B` be two square matrices of order n. By considering them as a sum of diagonals (i.e., for `A`: `A`$= \sum_{d=-(n-1)}^{n-1} A_d$ where A_d is a matrix with non-zero entries only on diagonal d), write the matrix multiplication `C=A*B` by using the previous function `mmultd`.

Exercise 1.12 *Accuracy of the inner product.*

Let $x, y \in \mathbb{R}^n$ and $z = x^T y$.

1. *Forward error in floating point arithmetic.* Complete the proof of identity (1.5)

2. *Condition number.* This number evaluates the sensitivity of a result with respect to perturbations of its operands. More precisely, let us consider two vectors of perturbations Δx and Δy such that they provoke an error Δa on the result: $a + \Delta a = (x + \Delta x)^T (y + \Delta y)$. Prove that:

$$\frac{|\Delta a|}{|a|} \leq C(x,y) \left(\frac{\|\Delta x\|_2}{\|x\|_2} + \frac{\|\Delta y\|_2}{\|y\|_2} \right) + \left(\frac{\|\Delta x\|_2 \, \|\Delta y\|_2}{\|x\|.\|y\|} \right), \quad (1.22)$$

where $C(x,y) = \frac{1}{\cos \angle(x,y)}$ is called the *condition number* of the computation. This results illustrates that the computation of the inner product of nearly orthogonal vectors is an ill-conditioned calculation.

3. *Backward error.* In floating point arithmetic, prove, from equation (1.5), that there exists $\Delta x \in \mathbb{R}^n$ such that

$$a_c = (x + \Delta x)^T y, \quad (1.23)$$

with $\|\Delta x\|_2 = \|x\|_2 \, O(n\epsilon_M)$. The computation is said to be *backward stable.*

4. *Condition number and Backward Error combines in the Forward Error.* From (1.22) and (1.23), prove that the relative forward error can be expressed by

$$\frac{|a_c - a|}{|a|} = O(C(x,y) \, n \, \epsilon_M). \quad (1.24)$$

Exercise 1.13 *Rounding errors in GAXPY operations.*

Complete the proof of (1.7).

Exercise 1.14 *Vandermonde matrices.*

Let $x \in \mathbb{R}^n$. Write a MATLAB program that generates:

$$V(x) = \begin{pmatrix} 1 & x_1 & x_1^2 & x_1^3 & \cdots & x_1^{n-1} \\ 1 & x_2 & x_2^2 & x_2^3 & \cdots & x_2^{n-1} \\ \vdots & \vdots & \vdots & \cdots & \cdots & \vdots \\ \vdots & \vdots & \vdots & \cdots & \cdots & \vdots \\ 1 & x_n & x_n^2 & x_n^3 & \cdots & x_n^{n-1} \end{pmatrix}$$

Compare $V(x)$ with the matrix obtained using the available MATLAB command vander.

Exercise 1.15 *Circulant matrices.*

Let $x \in \mathbb{R}^n$. Write a MATLAB program that generates:

$$C(x) = \begin{pmatrix} x_1 & x_2 & \cdots & x_j & \cdots & x_n \\ x_n & x_1 & . & x_{j-1} & \cdots & x_{n-1} \\ \cdots & \cdots & \cdots & \cdots & \cdots & \cdots \\ \cdots & \cdots & \cdots & \cdots & \cdots & \cdots \\ x_2 & x_3 & \cdots & \cdots & \cdots & x_1 \end{pmatrix}$$

Exercise 1.16 *Toeplitz matrices.*

Let $c \in \mathbb{R}^n$ be a column vector and $r \in \mathbb{R}^{1 \times n}$, ($r(1)$ unused) be a row vector. Write a MATLAB program that generates:

$$T(c,r) = \begin{pmatrix} c_1 & r_2 & \dots & \dots & r_{n-1} & r_n \\ c_2 & c_1 & \dots & \dots & \dots & r_{n-1} \\ \dots & \dots & \dots & \dots & \dots & \dots \\ c_{n-1} & \dots & \dots & \dots & c_1 & r_2 \\ c_n & c_{n-1} & \dots & \ .. & c_2 & c_1 \end{pmatrix}$$

Exercise 1.17

Check the matrices obtained using the commands: magic, pascal, hilb. Write the MATLAB functions mypascal(n) and myhilb(n).

Exercise 1.18

1. Complete the following MATLAB sparse matrix multiplication which computes $C = A \times B$ for 2 n by n sparse matrices, by computing as a scalar product the coefficient $C(i,j) = A(i,:) \star B(:,j)$ whenever that coefficient is no 0.

Algorithm 1.5 Sparse Matrix Multiplication

```
function C=SMatMult(A,B)
% Input: Sparse n by n matrices A,B
% Output: a possibly sparse matrix C
n=length(A);
nz=n*n;%maximum number of non zero elements in C
% Define the 3 COO column vectors of C
I=zeros(nz,1);J=zeros(nz,1);c=zeros(nz,1);
k=0;%points to entries of 1D storage of c
for j = 1:n
    x=            ;% Get column vector j in B
    Ix=           ;% with indices of non zero entries
    for i=1:n
        v=            ;% Get row vector i in A
        Iv=           ;% with indices of non zero entries
        Inz=intersect(Ix,Iv);
        %Returns in Iv indices of non-zero elements in row i
        if isempty(Inz)==0
            k=k+1;I(k)=        ;J(k)=        ;c(k)=        ;
        end
    end
end
Ic=find(I);C=                              ;
```

---

2. Using the `MATLAB` command `sprand` for matrices A, B, of size $n \in \{2^p \mid p = 4, 5, 6, 7, 8, 9, 10\}$ and density $d \in \{0.1, 0.2, 0.4\}$ test this program as follows:

   - By comparing the execution times of `SMatMult` against `MATLAB *`, using the functions `tic, toc`.

   - By finding the number of non-zero elements of C, using the `MATLAB` command `nnz(C)` as a function of the common density of A and B.

   - Using the `MATLAB` command `spy` find the profiles of the resulting matrices C.

Exercise 1.19

Implement the (CSR) type storage of a sparse matrix A, using the same procedure as that in the (CSC).

1.9 Computer Project: Strassen Algorithm

This algorithm replaces the block multiplication described in Section 1.6.4 by a procedure that involves less arithmetic operations.

Let us consider three square matrices A, B and C of order $n = 2m$ which are partitioned as in (1.12-1.14). Each block is supposed to be square of order m. The regular block multiplication is expressed in the formulae (1.15-1.18). It involves 4 additions and 8 multiplications of blocks of order m. In the Strassen multiplication, computations are rearranged into 7 multiplications and 18 additions of blocks:

$$
\begin{aligned}
P_1 &= (A_{11} + A_{22})(B_{11} + B_{22}), \\
P_2 &= (A_{21} + A_{22})B_{11}, \\
P_3 &= A_{11}(B_{12} - B_{22}), \\
P_4 &= A_{22}(B_{21} - B_{11}), \\
P_5 &= (A_{11} + A_{12})B_{22}, \\
P_6 &= (A_{21} - A_{11})(B_{11} + B_{12}), \\
P_7 &= (A_{12} - A_{22})(B_{21} + B_{22}), \\
C_{11} &= P_1 + P_4 - P_5 + P_7, \\
C_{12} &= P_3 + P_5, \\
C_{21} &= P_2 + P_4, \\
C_{12} &= P_1 + P_3 - P_2 + P_6.
\end{aligned}
$$

For the seven multiplications, the regular algorithm is used.

1. Prove that this leads correctly to the product $C = AB$.

2. Prove that the number of flops $\mathcal{C}_1(n)$ when the Strassen reduction is applied is

$$\mathcal{C}_1(n) = 2(7m^3) + 18m^2 = \frac{7}{8}(2n^3) + \frac{9}{2}n^2.$$

Compare it to the regular complexity $\mathcal{C}_0(n) = 2n^3$ of the matrix multiplication and determine the minimum n such that $\mathcal{C}_1(n) < \mathcal{C}_0(n)$.

3. Let us assume that $n = 2^k m$, and that k Strassen reductions are recursively applied by the procedure described by the algorithm 1.6. Justify the algorithm.

Algorithm 1.6 Strassen's Algorithm

```
function C = strassen(A,B,k)
%
% C = strassen(A,B,k)
%
% Matrix multiplication by Strassen's reduction C=A*B
% Recursive calls with a maximum of k reductions
% The matrices A and B are supposed to be square of the same
   order n
%
n=size(A,1);
if k == 0 | mod(n,2) ~= 0,
    C= A*B ;
else
    m=n/2 ; C=zeros(n) ;
    I=[1:m]; J=[m+1:n] ;
    P1 = strassen(A(I,I)+A(J,J),B(I,I)+B(J,J),k-1) ;
    P2 = strassen(A(J,I)+A(J,J),B(I,I),k-1) ;
    P3 = strassen(A(I,I),B(I,J)-B(J,J),k-1) ;
    P4 = strassen(A(J,J),B(J,I)-B(I,I),k-1) ;
    P5 = strassen(A(I,I)+A(I,J),B(J,J),k-1) ;
    P6 = strassen(A(J,I)-A(I,I),B(I,I)+B(I,J),k-1) ;
    P7 = strassen(A(I,J)-A(J,J),B(J,I)+B(J,J),k-1) ;
    C(I,I) = P1+P4-P5+P7 ;
    C(I,J) = P3+P5 ;
    C(J,I) = P2+P4 ;
    C(J,J) = P1+P3-P2+P6 ;
end;
```

---

4. When $n = 2^k$, prove that by applying k Strassen reductions, the total complexity is

$$\mathcal{C}_k(n) = 7n^{\log_2 7} - 6n^2.$$

Compare with $\mathcal{C}_0(n)$.

Chapter 2

Basic Concepts for Matrix Computations

In Chapter 1, we provided the basics of matrix algebra through the BLAS operations. In this chapter, we give additional notations and concepts used in matrix computations. For additional and detailed material on related topics, we refer the reader to the books by Strang [65], Ciarlet [19], Householder [35] and Golub and Van Loan [33].

We start by introducing topological structures on the finite dimensional vector spaces $\mathbb{R}^n$ and $\mathbb{C}^n$.

2.1 Vector Norms

Vector norms are defined on finite or infinite dimensional vector spaces. They extend the notion of absolute value (and modulus) on $\mathbb{R}$ (and $\mathbb{C}$). We proceed as follows.

Definition 2.1 *A vector norm on $\mathbb{K}^n$, $\mathbb{K} = \mathbb{R}$ or $\mathbb{K} = \mathbb{C}$, is a function $||.|| : \mathbb{K}^n \to [0, \infty)$, satisfying the following properties:*

1. $||x|| \geq 0, \forall x \in \mathbb{K}^n$ *and* $||x|| = 0 \Leftrightarrow x = 0$,

2. $||x + y|| \leq ||x|| + ||y||$, $\forall x, y \in \mathbb{K}^n$,

3. $||cx|| = |c|||x||, \forall c \in \mathbb{K}, \ \forall x \in \mathbb{K}^n$.

The l^p norms constitute an essential class of vector norms. These are defined as follows.

Definition 2.2 *The l^p norm of $x = (x_1, ..., x_n) \ \forall x \in \mathbb{K}^n$, is defined by:*

1. $||x||_p = (\sum_{i=1}^{n} |x_i|^p)^{1/p}, \forall p, 1 \leq p < \infty$

2. $||x||_\infty = \max_{1 \leq i \leq n} |x_i|$

For example:

$$||x||_1 = |x_1| + |x_2| + ... + |x_n|$$

For the case $p = 2$, recall from Chapter 1, the definition of the scalar product as a BLAS-1 operation:

$$\forall x, y \in \mathbb{R}^n : x^T y = \sum_{i=1}^{n} x_i y_i = y^T x,$$

and

$$\forall x, y \in \mathbb{C}^n : x^* y = \sum_{i=1}^{n} x_i \overline{y_i} = \overline{y^* x}.$$

Hence:

$$||x||_2 = (|x_1|^2 + |x_2|^2 + ... + |x_n|^2)^{1/2} = \begin{cases} (x^T x)^{1/2} \text{ for } x \in \mathbb{R}^n \\ (x^* x)^{1/2} \text{ for } x \in \mathbb{C}^n \end{cases}$$

Note the usual inequalities on l^p :
Hölder's inequality:

$$\forall x, y \in \mathbb{C}^n, |x^* y| \leq ||x||_p ||y||_q, \text{ if } \frac{1}{p} + \frac{1}{q} = 1$$

In the particular case when $p = q = 2$, one such inequality becomes the Cauchy-Schwarz inequality:

$$|x^* y| \leq ||x||_2 ||y||_2.$$

This last inequality can be obtained from the inequality:

$$||\alpha x + y||_2^2 = (\alpha x + y)^* (\alpha x + y) = |\alpha|^2 ||x||_2^2 + 2Re(\alpha x^* y) + ||y||_2^2 \geq 0.$$

Since, $2\Re e(\alpha x^* y) \leq |\alpha| |x^* y|$, we get:

$$|\alpha|^2 ||x||_2^2 + 2|\alpha| |x^* y| + ||y||_2^2 \geq ||\alpha x + y||_2^2 \geq 0, \forall x, y \in \mathbb{C}^n, \forall \alpha \in \mathbb{C}.$$

This is possible if and only if:

$$|x^* y|^2 - ||x||_2 ||y||_2 \leq 0.$$

The equality occurs if and only if $x + \alpha y = 0$, that is if x and y are co-linear. Some vector norms are implemented in MATLAB:

```
norm(x,p) = sum(abs(x).^p)^(1/p).
norm(x) = norm(x,2)=sqrt(x'*x).
norm(x,inf) = max(abs(x)).
```

2.2   Complements on Square Matrices

We recall first a fundamental theorem from linear algebra related to square matrices and associated systems of linear equations.

Theorem 2.1 *For a matrix $A \in \mathbb{C}^{n \times n}$, one has the following equivalences:*

1. *The matrix A is invertible, i.e., there exists a matrix $B = A^{-1} \in \mathbb{C}^{n \times n}$, the inverse matrix of A, such that $A \times B = B \times A = I$, I being the identity matrix on $\mathbb{C}^n$.*

2. *For $b \in \mathbb{C}^n$, the system $Ax = b$ admits a unique solution $x \in \mathbb{C}^n$.*

Numerical algorithms for finding matrix inverses will be detailed in Chapter 3 of this book.

2.2.1 Definition of Important Square Matrices

Recall the n by n identity matrix on $\mathbb{R}^n$ (or $\mathbb{C}^n$), given by:

$$I_n = [e_1, \, e_2, \, ..., e_{n-1}, \, e_n],$$

where each vector e_i has all its components equal to 0 except the ith component equal to 1.

We denote I_n by I when no ambiguity exists about the matrix order. The basis $\{e_1, e_2, ..., e_n\}$ is referred to as the canonical basis in $\mathbb{R}^n$ and $\mathbb{C}^n$.

In Table 2.1, we provide the case of some important square matrices and, correspondingly, their `MATLAB` representation.

Case	MATLAB **justification**
Identity n by n matrix: $I_n \equiv I$	`eye(n)`
Inverse matrix of $A \in \mathbb{C}^{n \times n}$: $A \times A^{-1} = I$	`inv(A)`
Symmetric matrix $A \in \mathbb{R}^{n \times n}$: $A = A^T$	`A==A'`
Hermitian matrix $A \in \mathbb{C}^{n \times n}$: $A = A^*$	`A==A'`
Orthogonal matrix $A \in \mathbb{R}^{n \times n}$: $A^T = A^{-1}$	`inv(A)==A'`
Unitary matrix $U \in \mathbb{C}^{n \times n}$: $U^* = A^{-1}$	`inv(U)==U'`

TABLE 2.1: `MATLAB` notations for some square matrices.

2.2.2 Use of Orthonormal Bases

We start by defining orthogonality in $\mathbb{R}^n$ and $\mathbb{C}^n$.

Definition 2.3 Vector Orthogonality
Two non-zero vectors x and y in $\mathbb{R}^n$ are said to be orthogonal if $x^T y = 0$.
Two non-zero vectors x and y in $\mathbb{C}^n$ are said to be orthogonal if $x^* y = 0$.

Note that $e_i^T e_j = e_i^* e_j = 0$, $i \neq j$ and $e_i^T e_i = e_i^* e_i = 1, \forall i$.
Also for $x \in \mathbb{R}^n$, one has:

$$x = (x^T e_1)e_1 + (x^T e_2)e_2 + ...(x^T e_i)e_i + ... + (x^T e_n)e_n,$$

and as well for $x \in \mathbb{C}^n$, one has:

$$x = (x^* e_1)e_1 + (x^* e_2)e_2 + ...(x^* e_i)e_i + ... + (x^* e_n)e_n.$$

More generally, we introduce **orthonormal bases**.

Definition 2.4 Orthonormal Bases
$F = [f_1, f_2, ..., f_n] \in \mathbb{R}^{n \times n}$ $(f_j = F(:,j))$ *is an orthonormal basis in* $\mathbb{R}^n$*, if*
$F^T F = I$.
Also, $F = [f_1, f_2, ..., f_n] \in \mathbb{C}^{n \times n}$ *is an orthonormal basis in* $\mathbb{C}^n$*, if* $F^* F = I$.

Note that if the column vectors of the square matrix $F = [f_1, f_2, ..., f_n]$ form an orthonormal basis, then F is an orthogonal matrix $(F \times F^T = F^T \times F = I)$, whenever the coefficients of F are real and, is unitary $(F \times F^* = F^* \times F = I)$ if those coefficients are complex.
In what follows *we restrict ourselves to real coefficient matrices unless complex coefficients matrices are needed.* In that case, this will be explicitly indicated. Consider $x \in \mathbb{R}^n$ and let $V = [v_1, v_2, ..., v_n]$ be any basis in $\mathbb{R}^n$. The column vector α of components of x,

$$\alpha = [\alpha_1, \alpha_2, \alpha_n]^T,$$

satisfy:

$$x = \alpha_1 v_1 + \alpha_2 v_2 + ... + \alpha_n v_n \iff x = V\alpha \iff \alpha = V^{-1} x.$$

In case $V \equiv F = [f_1, f_2, ..., f_n]$ is an orthonormal basis, then the components of x with respect to F are given by the column vector $\alpha = F^T x$.

2.2.3   Gram-Schmidt Process

We take now a first look at the Gram-Schmidt process which transforms any set of m linearly independent vectors $V = [v_1, v_2, ..., v_m]$ in $\mathbb{R}^n$, $m \leq n$[1] into an orthonormal set $F = [f_1, f_2, ..., f_m]$ of m linearly independent vectors. The generalization of this process to the complex case is done in Computer Exercise 2.1.
One needs first some preliminary notations.

Definition 2.5 *Let* $\mathcal{U}$ *be a subspace of* $\mathbb{R}^n$ *such that* $\mathcal{U}$ *includes at least one vector* $u \neq 0$ $(\dim(\mathcal{U}) \geq 1)$*. The orthogonal subspace* $\mathcal{U}^\perp$ *of* $\mathcal{U}$ *is defined by:*

$$\mathcal{U}^\perp = \{x \in \mathbb{R}^n \,|\, x^T u = 0, \forall u \in \mathcal{U}\}.$$

We turn now to the general definition of *orthogonal projections*.

---

[1]$V \in \mathbb{R}^{m \times n}$, $m \leq n$ is said to be a full-rank system of vectors.

Definition 2.6 *Let $\mathcal{U}$ be a subspace of $\mathbb{R}^n$ such that $\mathcal{U}$ includes at least one vector $u \neq 0$ $(\dim(\mathcal{U}) \geq 1)$. The orthogonal projection $\mathcal{P}_{\mathcal{U}}$ from $\mathbb{R}^n$ to $\mathcal{U}$, $(\mathcal{P}_{\mathcal{U}} : \mathbb{R}^n \rightarrow \mathcal{U})$ is defined by:*

$$\forall x \in \mathbb{R}^n : v^T(x - \mathcal{P}_{\mathcal{U}}x) = 0, \forall v \in \mathcal{U} \text{ and } x \in \mathcal{U} \Longrightarrow \mathcal{P}_{\mathcal{U}}x = x$$

It is well known that any vector $x \in \mathbb{R}^n$ can be decomposed uniquely into $\mathcal{U}$ and $\mathcal{U}^\perp$, i.e.,

$$x = x_{\mathcal{U}} + x_{\mathcal{U}^\perp},$$

where $x_{\mathcal{U}} = \mathcal{P}_{\mathcal{U}}x$ and $x_{\mathcal{U}^\perp} = x - x_{\mathcal{U}} = \mathcal{P}_{\mathcal{U}^T}x$ are, respectively, the *orthogonal projections* of x on $\mathcal{U}$ and $\mathcal{U}^\perp$. One consequently writes:

$$\mathbb{R}^n = \mathcal{U} \oplus \mathcal{U}^\perp \qquad (2.1)$$

i.e., $\mathbb{R}^n$ is the direct sum of $\mathcal{U}$ and $\mathcal{U}^\perp$.

If for $k \geq 1$, we let $\mathcal{U}_k = span\{f_1, f_2, ..., f_k\}$, where $\{f_1, f_2, ..., f_k\}$ is an orthonormal basis in $\mathcal{U}_k$, then by defining the matrix:

$$F_k = [f_1, \ f_2, \ ..., \ f_k],$$

then one has:

$$x_{\mathcal{U}_k} = \mathcal{P}_{\mathcal{U}_k}(x) = \Sigma_{i=1}^k (x^T f_i)f_i = F_k(F_k^T x) = F_k F_k^T x,$$

and

$$x - x_{\mathcal{U}_k} = (I - F_k F_k^T)x.$$

We state now the Gram-Schmidt procedure.

Theorem 2.2 Gram-Schmidt Process for Full-Rank Systems *Let $V = [v_1, v_2, ..., v_m] \in \mathbb{R}^{n \times m}$, $m \leq n$ be a full rank system $\operatorname{rank}(V) = m$ and let:*

$$V = span\{v_1, v_2, ..., v_m\}$$

Then $\mathcal{V} = span\{f_1, f_2, ..., f_m\}$ where $F = [f_1, f_2, ..., f_m]$ is an orthonormal basis, i.e., $F^T F = I_m$.

Proof. This result follows directly from the recurrence process:

1. Base Step: Start with v_1. Let $f_1 = \frac{1}{c_1}v_1$ with c_1 chosen such that $||f_1||^2 = (f_1)^T f_1 = 1$.

2. Iterative Step: $1 < k \leq m$, consider the subspace $\mathcal{U}_k = span(f_1, ..., f_{k-1})$ and let then:

$$w_k = v_k - \mathcal{P}_{\mathcal{U}_k}(v_k) = v_k - (v_k^T f_1)f_1 - ...(v_k^T f_{k-1})f_{k-1}.$$

Letting $F_k = [f_1, f_2, ..., f_k]$, $k \geq 1$, then $\mathcal{P}_{\mathcal{U}_k}(v_k) = F_k F_k^T v_k$ and

$$w_k = v_k - F_k F_k^T v_k = (I - F_k F_k^T)v_k.$$

It is easily shown that $w_k \neq 0$ and is orthogonal to $\mathcal{U}_k$. Choose then c_k and $f_k = \frac{1}{c_k}w_k$, such that $||f_k||^2 = (f_k)^T f_k = 1$.

Consequently, by induction, one proves that such process leads to a new orthonormal set $\{f_i | i = 1, ..., n\}$, i.e.,

$$(f_i)^T(f_j) = 0, \ i \neq j \ \text{and} \ (f_i)^T(f_i) = 1, \forall i,$$

in which one obtains $[f_1, f_2, ..., f_m]$ from $[v_1, v_2, ..., v_m]$ using the relations:

$$
\begin{aligned}
v_1 &= c_1 f_1 \\
v_2 &= c_2 f_2 + (v_2^T f_1) f_1 \\
&\ldots\ldots \\
v_i &= c_i f_i + (v_i^T f_1) f_1 + ... + (v_i^T f_{i-1}) f_{i-1} \\
&\ldots\ldots \\
v_m &= c_m f_n + (v_m^T f_1) f_1 + ... + (v_m^T f_{n-1}) f_{m-1}.
\end{aligned}
\tag{2.2}
$$

Since $c_i \neq 0$, $i = 1, ... m$ then $span(F) = span(V) = \mathcal{V}$ ∎

Figure 2.1 shows the first steps of a Gram-Schmidt process transforming a linearly independent set $\{U_1, U_2\}$ into an orthonormal set $\{v_1, v_2\}$.

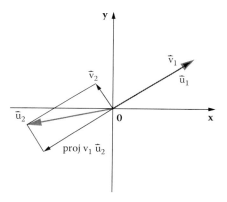

FIGURE 2.1: A Gram-Schmidt process transforming $\{u_1, u_2\}$ into $\{v_1, v_2\}$

Algorithm 2.1 is an implementation of the Gram-Schmidt process as demonstrated in Theorem 2.2.

Algorithm 2.1 Gram-Schmidt Orthogonalization

```
function F=GramSchmidt(V,tol)
% Input: V an n by m matrix m<=n
%        tol: tolerance to measure ill-conditioning of V
% Output: F an n by m orthogonal matrix
[n,m]=size(V);F=zeros(n,m);
f=(1/a)*e;F(:,1)=f;% n flops
for k=2:n
    e=V(:,k);    % Project v(:,k) on span{F(:,1),...,F(:,k-1)}
    Fk=F(:,1:k-1);vc=(e'*Fk)';%2(k-1)n flops
    w=e-Fk*vc;a=norm(w);%2n*(k-1)+3n flops
    if a<tol disp(['Process breaks at:' num2str(k) ',
    rank(V) <n']);
    break end
    F(:,k)=(1/a)*w;%n flops
end
```

Remark 2.1 *Note that Theorem 2.2 can be generalized to a rank deficient matrix $V = [v_1, v_2, ..., v_m]$ of m (not necessarily linearly independent) vectors in $\mathbb{R}^n$, $m \leq n$. In that case, the Gram-Schmidt process allows also to determine the rank of V as defined in Section 2.3.*

This issue is left to Exercise 2.1. More generally, the Gram-Schmidt procedure on any set V of n column vectors in $\mathbb{R}^m$ will be treated in Chapter 4 as a one of the orthogonal methods that would lead to a QR decomposition of V.

We turn now to the definitions on determinants and eigenvalues.

2.2.4 Determinants

Definition 2.7 *Given a square matrix $A \in \mathbb{K}^n$, the determinant of A is given by:*

$$\det(A) = \sum_{(i_1 i_2 \ldots i_n) \in P(1\,2\,\ldots n)} (-1)^p a_{i_1 1} a_{i_2 2} \ldots a_{i_n n}, \qquad (2.3)$$

where $P(1\,2\,...n)$ is the set of all $(n!)$ permutations on the indices $(1\,2\,...n)$ and $(-1)^p$ is the signature of $(i_1 i_2 \ldots i_n)$, p being the number of transpositions needed to transform $(i_1 i_2 \ldots i_n)$ into $(1\,2\,...n)$.

One has the following determinant properties:

1. Let $A \in \mathbb{K}^{n \times n}$. The determinant is a function det: $A \in \mathbb{K}^{n \times n} \to \mathbb{K}$ that is homomorphic with respect to the multiplication of square matrices, i.e.,

$$\det(A \times B) = \det(A) \times \det(B), \forall A, B \in \mathbb{C}^{n \times n}, \qquad (2.4)$$

2. For the identity matrix I in $\mathbb{K}^{n \times n}$, i.e., $\det(I) = 1$.

3. The last 2 properties imply that

$$\det(A \times A^{-1}) = \det(A)\det(A^{-1}) = 1.$$

4. Therefore, a square matrix A is invertible **if and only if** $\det(A) \neq 0$, implying in that case that

$$\det(A^{-1}) = \frac{1}{\det(A)}.$$

This characterizes the solvability of a system of n linear equations in n unknown. Specifically one has:

Theorem 2.3 *For a matrix $A \in \mathbb{K}^{n \times n}$ and $b \in \mathbb{K}^n$ the system $Ax = b$ admits a unique solution if and only if $\det(A) \neq 0$.*

Note that using (2.3), the computation of a determinant requires **at most** $n(n!)$ multiplications and $n!$ additions. One way to reduce the number of operations is discussed in Chapter 3, whereas the evaluation of the determinant corresponds to that of a triangular in which case, it requires only n multiplications. Note also the following properties.

1. $\det(A^T) = \det(A)$

2. $\det(cA) = c^n \det(A)$

3. If A is upper triangular, then $\det(A) = a_{11}a_{22}...a_{nn}$.

2.2.5 Eigenvalue-Eigenvector and Characteristic Polynomial

Pair Eigenvalue-Eigenvector

Definition 2.8 *Let $A \in \mathbb{C}^{n \times n}$ be a square matrix. The pair $\{\lambda \in \mathbb{C}, x_\lambda \in \mathbb{C}^n\}$ is said to be an **eigenpair** of **eigenvalue - eigenvector** for A if they satisfy:*

$$x_\lambda \neq 0, \ Ax_\lambda = \lambda x_\lambda. \tag{2.5}$$

The set of all repeated eigenvalues of the matrix A is called the **spectrum** of A and is denoted by $\Lambda(A)$.

Characteristic Polynomial

Proposition 2.1 $\lambda \in \Lambda(A)$ *if and only if:*

$$\det(A - \lambda I) = 0. \tag{2.6}$$

(I being the identity matrix).

Proof. Note that $\lambda \in \Lambda(A)$ is equivalent to (2.5), which in turn is equivalent to $A - \lambda I$ not invertible, i.e., $\det(A - \lambda I) = 0$. ∎

Let $p_A(\lambda) = \det(A - \lambda I)$. Note that $p_A(\lambda)$ is a polynomial of degree exactly n (as can be shown in Exercise 2.8), which has real or complex coefficients depending on whether $A \in \mathbb{R}^{n \times n}$ or $A \in \mathbb{C}^{n \times n}$.
From the fundamental theorem of algebra regarding polynomials of degree on $\mathbb{C}$, one may assert the following:

Proposition 2.2 *For a square n by n matrix A with real or complex coefficients:*

    1. A has at most n eigenvalues, distinct or repeated in $\mathbb{C}$.

    2. If A has real coefficients then, $\lambda \in \mathbb{C} \in \Lambda(A)$ implies $\overline{\lambda} \in \Lambda(A)$.

As a consequence one can characterize invertibility of square matrices, through their spectrum.

Proposition 2.3 *A square matrix A is invertible (non-singular) if and only if $0 \notin \Lambda(A)$.*

Proof. It follows from $0 \in \Lambda(A)$ if and only if $\det(A - 0I) = \det(A) = 0$. ∎

The **spectral radius** of a square matrix A, $\rho(A)$, is the largest modulus of the elements of the spectrum of A, i.e.,

$$\rho(A) = \max_{\lambda \in \Lambda(A)} |\lambda|.$$

In MATLAB there are 2 commands that solve the eigenvalue problem:
1. **function eig**:

```
E = eig(A)
    gives a vector containing all the eigenvalues of a square
    matrix A.
    [V,D] = eig(A)
    is the syntax of this command.
    It produces a diagonal matrix D of eigenvalues
    and a full matrix V whose columns are the eigenvectors
    corresponding to D, so that
    A*V=V*D (A=V*D*inv(V))
```

and

2. **function eigs**

```
D = eigs(A)
    returns a vector of A's 6 largest magnitude eigenvalues.
    A must be square and should be large and sparse,
    though eigs will work on full matrices as well.
```

```
[V,D] = eigs(A) returns a diagonal matrix D of A's
6 largest magnitude eigenvalues
and a matrix V whose columns are the corresponding
eigenvectors.
eigs(A,k) returns the k largest magnitude eigenvalues.
eigs(A,k,sigma) returns k eigenvalues:
        the largest in magnitude if sigma is 'LM'
        the smallest magnitude if sigma is 'SM'
```

The number of linearly independent eigenvectors (eigenspace) associated with an eigenvalue λ_0 is its *geometric multiplicity*, $m_g(\lambda_0)$, while the algebraic multiplicity of λ_0, $m_a = m_a(\lambda_0)$ is its multiplicity as a root of $p_A(\lambda)$, that is the largest integer m_a such that $p_A(\lambda) = (\lambda - \lambda_0)^{m_a} q_A(\lambda)$, with:

$$\text{degree}(q_A) = n - m_a.$$

Note also that $1 \leq m_g \leq m_a \leq n$. This is well illustrated in the example of Exercise 2.2.

Chapter 5 of this book is devoted to the study of algorithms for finding eigenvalues for matrices. However, for the purpose of "practicing" the concepts presented in this chapter, the reader will be invited to construct algorithms based on one of `eig` or `eigs`. This is in particular the case for Schur's decomposition of square matrices.

2.2.6   Schur's Decomposition

We give now Schur's decomposition theorem, which is a fundamental result for any square matrix.

Theorem 2.4 *Let* $A \in \mathbb{C}^{n \times n}$. *There exists an upper triangular matrix* $T \in \mathbb{C}^{n \times n}$ *and a unitary matrix* $U \in \mathbb{C}^{n \times n}$ $(U^*U = UU^* = I)$, *such that*

$$A = UTU^* \ i.e., \ AU = UT, \tag{2.7}$$

i.e., the triangular matrix T *is "similar"[2] to* A.

Proof. The proof is done by recurrence starting with the base step in which we pick $\lambda_1 \in \mathbb{C}$ as one eigenvalue of A with corresponding unit eigenvector $f_1 \in \mathbb{C}^n$ (i.e., $f_1^* f_1 = 1$). For simplicity, λ_1 may be an eigenvalue of A with smallest or largest modulus. Let also $\mathcal{U}_1$ be the orthogonal subspace to f_1; $\dim(\mathcal{U}_1) = n - 1$. Using the Gram-Schmidt process, one constructs an orthonormal basis $\{f_2^1, ..., f_n^1\}$ in $\mathcal{U}_1$. Since $A : \mathbb{C}^n \to \mathbb{C}^n$ is a linear application, then the matrix $A^{(1)}$ that expresses such application with respect to $F^{(1)} = \{f_1, f_2^1, ..., f_n^1\}$ satisfies:

$$AV^{(1)} = V^{(1)}A^{(1)}, \ (A^{(1)} \text{ is similar to } A). \tag{2.8}$$

---

[2]Two square matrices B and C are said to be similar if there exists an invertible matrix P, such that $BP = PC$ $(B = PCP^{-1})$. Consequently, $\det(B) = \det(C)$ and $\Lambda(B) = \Lambda(C)$.

with $V^{(1)}$ being the unitary matrix representing the base $\{f_1, f_2^1, ..., f_n^1\}$. As $A^{(1)} f_1 = \lambda_1 f_1$, then the first column of the matrix $A^{(1)}$ is $(\overline{\lambda_1} \, 0 .... 0)^*$ and $A^{(1)}$ has the following form:

$$A^{(1)} = \begin{pmatrix} \lambda_1 & A_{12}^{(1)} \\ O_{n-1,1} & A_{22}^{(1)} \end{pmatrix}$$

where $O_{n-1,1}$ is a zero column vector of size $n-1$, $A_{12}^{(1)} \in \mathbb{C}^{1\times(n-1)}$ and $A_{22}^{(1)} \in \mathbb{C}^{(n-1)\times(n-1)}$.

From (2.8), one has $A = V^{(1)} A^{(1)} (V^{(1)})^*$ and $A - \lambda I = V^{(1)} (A^{(1)} - \lambda I)(V^{(1)})^*$. So that

$$p_A(\lambda) = p_{A^{(1)}}(\lambda),$$

and therefore, $\Lambda(A^{(1)}) = \Lambda(A)$.

We may then start the recurrence process and consider the eigenvalues of $A_{22}^{(1)}$, given that

$$p_{A^{(1)}}(\lambda) = (\lambda - \lambda_1) p_{A_{22}^{(1)}}(\lambda),$$

and therefore,

$$\Lambda(A_{22}^{(1)}) \subseteq \Lambda(A),$$

with $\Lambda(A_{22}^{(1)}) = \Lambda(A) - \{\lambda_1\}$ if and only if λ_1 has an (algebraic) multiplicity equal to 1.

Picking λ_2 as an eigenvalue of $A_{22}^{(1)}$ (along the same way as λ_1 was selected for A in the base step), we show that there exist an orthogonal matrix $V_{22}^{(2)} \in \mathbb{C}^{(n-1)\times(n-1)}$, such that

$$A_{22}^{(1)} V_{22}^{(2)} = V_{22}^{(2)} \begin{pmatrix} \lambda_2 & A_{12}^{(2)} \\ O_{n-2,1} & A_{22}^{(2)} \end{pmatrix}$$

where $O_{n-2,1}$ is a zero column vector of size $n-2$, $A_{12}^{(2)} \in \mathbb{C}^{1\times(n-2)}$ and $A_{22}^{(2)} \in \mathbb{C}^{(n-2)\times(n-2)}$. If we set:

$$V^{(2)} = \begin{pmatrix} 1 & (O_{n-1,1})^* \\ O_{n-1,1} & V_{22}^{(2)} \end{pmatrix},$$

then:

$$AV^{(1)}V^{(2)} = V^{(1)}A^{(1)}V^{(2)} = V^{(1)} \begin{pmatrix} \lambda_1 & A_{12}^{(1)} \\ O_{n-1,1} & A_{22}^{(1)} \end{pmatrix} \begin{pmatrix} \lambda_1 & (O_{n-1,1})^* \\ O_{n-1,1} & V_{22}^{(2)} \end{pmatrix}$$

i.e.,

$$AV^{(1)}V^{(2)} = V^{(1)}A^{(1)}V^{(2)} = V^{(1)} \begin{pmatrix} \lambda_1 & A_{12}^{(1)}V_{22}^{(2)} \\ O_{n-1,1} & A_{22}^{(1)}V_{22}^{(2)} \end{pmatrix} = ...$$

$$....V^{(1)}\begin{pmatrix} \lambda_1 & A_{12}^{(1)}V_{22}^{(2)} \\ O_{n-1,1} & V_{22}^{(2)}A_{22}^{(2)} \end{pmatrix}$$

Thus,

$$AV^{(1)}V^{(2)} = V^{(1)}V^{(2)}\begin{pmatrix} \lambda_1 & A_{12}^{(1)}V_{22}^{(2)} \\ O_{n-1,1} & A_{22}^{(2)} \end{pmatrix}$$

i.e.,

$$AV^{(1)}V^{(2)} = V^{(1)}V^{(2)}\begin{pmatrix} \lambda_1 & A_{12}^{(1)}V_{22}^{(2)} \\ O_{n-1,1} & \begin{pmatrix} \lambda_2 & A_{12}^{(2)} \\ O_{n-2,1} & A_{22}^{(2)} \end{pmatrix} \end{pmatrix} = ...$$

$$...V^{(1)}V^{(2)}\begin{pmatrix} T_{11}^{(2)} & A_{12}^{(3)} \\ O_{n-2,2} & A_{22}^{(2)} \end{pmatrix}$$

with

$$T_{11}^{(2)} = \begin{pmatrix} \lambda_1 & t_{12} \\ 0 & \lambda_2 \end{pmatrix} \in \mathbb{C}^{2\times2}$$

$O_{n-2,2}$ an $n-2$ by 2 zero matrix and $A_{12}^{(2)} \in \mathbb{C}^{2\times(n-2)}$. Letting:

$$A^{(2)} = \begin{pmatrix} T_{11}^{(2)} & A_{12}^{(3)} \\ O_{n-2,2} & A_{22}^{(2)} \end{pmatrix}$$

one has after the second step:

$$AV^{(1)}V^{(2)} = V^{(2)}V^{(1)}A^{(2)}. \tag{2.9}$$

By induction, the process is continued, for $k = 3, ..., n-1$, thus generating the unitary matrices:

$$\{V^{(1)}, V^{(2)}, ..., V^{(n-1)}\}$$

until we reach the upper triangular matrix:

$$T = A^{(n)} = \begin{pmatrix} \lambda_1 & - & - & \cdots & - \\ 0 & \lambda_2 & - & \cdots & - \\ \vdots & 0 & \ddots & \lambda_{n-1} & - \\ 0 & 0 & 0 & 0 & \lambda_n \end{pmatrix}$$

such that

$$AV^{(1)}V^{(2)}...V^{(n-1)} = V^{(1)}V^{(2)}...V^{(n-1)}T.$$

Letting $U = V^{(1)}V^{(2)}...V^{(n-1)}$, we check that U is unitary, leading to Schur's decomposition.   ■

Corollary 2.1 *The set of eigenvalues of $A \in \mathbb{C}^{n\times n}$ is exactly the set of the diagonal elements of the matrix T (or D when A is symmetric), in the Schur decomposition of A.*

Proof. The proof is based on (2.7) and uses the following identity:

$$A - \lambda I = U^*(T - \lambda I)U,$$

so that

$$\det(A - \lambda I) = \det(U^*)\det(T - \lambda I)\det(U).$$

Since $\det(U^*)\det(U) = \det(U^*U) = 1$, then:

$$\det(A - \lambda I) = \det(T - \lambda I) = \Pi_{i=1}^{n}(T_{ii} - \lambda).$$

Thus,

$$\Lambda(A) = \{T_{ii} : i = 1, ..., n\}.$$

∎

In Exercise 2.2, we construct and analyze an algorithm that illustrates the proof of this theorem.

Schur's decomposition is implemented in `MATLAB` using the QR algorithm (see Chapter 5 on the eigenvalue problem.)

```
[U,T] = schur(A) for A square matrix acts as follows:
  If A is complex, the complex Schur form is returned in matrix T,
  an upper triangular with the eigenvalues of X on the diagonal.
  with A = U*T*U' and U'*U = eye(size(U)).
If A is real, two different decompositions are available.
 schur(A,'real'), the default is quasi-triangular.
 It has the real eigenvalues on the diagonal
 and the complex eigenvalues in 2-by-2 blocks on the diagonal.
 schur(A,'complex') is triangular.
 It results as if A is complex.
```

2.2.7 Orthogonal Decomposition of Symmetric Real and Complex Hermitian Matrices

Consider now the cases of a symmetric matrix $A \in \mathbb{R}^{n \times n}$, $A^T = A$ and also that of a Hermitian matrix $A \in \mathbb{C}^{n \times n}$, $A^* = A$.
We intend to apply Schur's decomposition theorem (2.7) for these 2 cases.

2.2.7.1   *A* **Real and Symmetric:** $A = A^T$

In such case $AU = UT$ leads to $T = U^*AU$ and $T^* = U^*A^*U = U^*A^TU$.
Thus, $T = T^*$, i.e., $T = D$, where D is a real diagonal matrix, i.e., the eigenvalues of A are real, and given that an eigenvector x corresponding to an eigenvalue λ satisfies $(A - \lambda I)x = 0$, hence $x \in \mathbb{R}^n$ and the matrix U of eigenvectors of A is also a real matrix. We denote it by $Q = U \in \mathbb{R}^{n \times n}$. Thus, one obtains the following theorem.

Theorem 2.5 *If $A \in \mathbb{R}^{n \times n}$ is symmetric, then:*
1. There exists an orthogonal matrix $Q = [q_1, q_2, ..., q_n] \in \mathbb{R}^{n \times n}$ of the eigenvectors of A and a diagonal matrix of the corresponding real eigenvalues of A:

$$D = diag \begin{pmatrix} \lambda_1 \\ \lambda_2 \\ \vdots \\ \lambda_{n-1} \\ \lambda_n \end{pmatrix},$$

such that

$$A = QDQ^T \iff AQ = QD \iff Aq_i = \lambda_i q_i, \ i = 1, ..., n.$$

2. Furthermore, $QQ^T = Q^T Q = I$, imply that the column vectors of Q form an orthonormal basis of eigenvectors of A in $\mathbb{R}^n$.

2.2.7.2 *A* Complex Hermitian: $A = A^*$

In such case $AU = UT$ gives $T = U^* AU$ and $T^* = U^* A^* U$. Thus, one has $T = T^* = D$, which makes also D a real diagonal matrix. Hence:

$$AU = UD. \tag{2.10}$$

and one obtains the equivalent of Theorem 2.5 for Hermitian matrices.

Theorem 2.6 *If $A \in \mathbb{C}^{n \times n}$ is Hermitian then:*
1. There exists a unitary matrix $U = [u_1, u_2, ..., u_n] \in \mathbb{R}^{n \times n}$ of the eigenvectors of A and a diagonal matrix of the corresponding real eigenvalues of A:

$$D = diag \begin{pmatrix} \lambda_1 \\ \lambda_2 \\ \vdots \\ \lambda_{n-1} \\ \lambda_n \end{pmatrix},$$

such that

$$A = UDU^* \iff AU = UD \iff Au_i = \lambda_i u_i, \ i = 1, ..., n.$$

2. Furthermore, $UU^* = U^* U = I$, imply that the column vectors of U form an orthonormal basis of eigenvectors of A in $\mathbb{C}^n$.

Consequently, when a square matrix A is either real symmetric or complex Hermitian, then its spectrum:

$$\Lambda(A) = \{\lambda_1, \lambda_2, ..., \lambda_n, \} \subset \mathbb{R},$$

which implies that in such cases, the eigenvalues can be indexed by increasing or decreasing order, i.e.,

$$\lambda_1 \leq \lambda_2 \leq ... \leq \lambda_n, \tag{2.11}$$

or

$$\lambda_1 \geq \lambda_2 \geq ... \geq \lambda_n. \tag{2.12}$$

2.2.8 Symmetric Positive Definite and Positive Semi-Definite Matrices

Definition 2.9 *A matrix $A \in \mathbb{R}^{n \times n}$ is said to be:*

1. **Positive definite (pd)** *if $\forall x \in \mathbb{R}^n$*
   $x^T A x \geq 0$, *with $x^T A x = 0$ if and only if $x = 0$.*

2. **Positive semi-definite** *if $\forall x \in \mathbb{R}^n$, $x^T A x \geq 0$.*

Definition 2.10 *A matrix $A \in \mathbb{R}^{n \times n}$ is said to be:*

1. **Symmetric positive definite (spd)** *if $A = A^T$ and $\forall x \in \mathbb{R}^n$*
   $x^T A x \geq 0$, *with $x^T A x = 0$ if and only if $x = 0$.*

2. **Symmetric positive semi-definite (spsd)** *if $A = A^T$ and $\forall x \in \mathbb{R}^n$*
   $x^T A x \geq 0$.

Definition 2.11 *A matrix $A \in \mathbb{C}^{n \times n}$ is said to be:*

1. **Positive definite (pd)** *if $\forall z \in \mathbb{C}^n$,*
   $Re\{z^* A z\} \geq 0$, *with $Re\{z^* A z\} = 0$ if and only if $z = 0$.*

2. **Positive semi-definite** *if $\forall z \in \mathbb{C}^n$, $Re\{z^* A z\} \geq 0$.*

Definition 2.12 *A matrix $A \in \mathbb{C}^{n \times n}$ is said to be:*

1. **Hermitian positive definite (hpd)** *if $A = A^*$ and $\forall z \in \mathbb{C}^n$*
   $z^* A z \geq 0$, *with $z^* A z = 0$ if and only if $z = 0$.*

2. **Hermitian positive semi-definite (hpsd)** *if $A = A^*$ and $\forall z \in \mathbb{C}^n$*
   $z^* A z \geq 0$.

As a consequence of Proposition 2.11, spd and spsd for real symmetric matrices and hpd and hpsd for complex Hermitian matrices have the following properties:

Proposition 2.4 *If a matrix $A \in \mathbb{R}^{n \times n}$ (or $A \in \mathbb{C}^{n \times n}$) is:*

1. *spd (hpd) then $\Lambda(A) \subset \mathbb{R}_+ = (0, \infty)$,*

2. *spsd (hpsd), then $\Lambda(A) \subset [0, \infty)$.*

Proof. Left for Exercises 2.9 and 2.10. ∎

2.3   Rectangular Matrices: Ranks and Singular Values

Rectangular matrices do not have eigenvalues, but do have *singular values*. For that purpose, we start by introducing the concept of matrix rank. Specifically, given a matrix $A \in \mathbb{C}^{m \times n}$, one may consider A as a mapping from $\mathbb{C}^n$ to $\mathbb{C}^m$, as:

$$x \in \mathbb{C}^n : y = Ax \in \mathbb{C}^m.$$

Hence, we can start by stating some basic concepts from Linear Algebra.

1. Range$(A) = \{y = Ax \in \mathbb{C}^m | x \in \mathbb{C}^n\}$, implying that $\mathbb{C}^m = \text{Range}(A) \oplus (\text{Range}(A))^{\perp}$.

2. Null$(A) = \{x \in \mathbb{C}^n | Ax = 0\}$, implying Null$(A) \neq \{0\}$ that $\mathbb{C}^n = \text{Null}(A) \oplus (\text{Null}(A))^{\perp}$

One has the following property:

Proposition 2.5 *Let A^* be the adjoint of $A \in \mathbb{C}^{m \times n}$ ($A^* = \overline{A}^T$). Then:*

1. $(Range(A))^{\perp} = Null(A^*)$

2. $(Range(A^*))^{\perp} = Null(A)$

Proof. Without loss in generality, we prove the second part. $\forall x \in \text{Null}(A) \subseteq \mathbb{C}^n$, $Ax = 0 \in \mathbb{C}^m$. Hence

$$\forall y \in \mathbb{C}^m, \forall x \in \text{Null}(A), \ y^* Ax = 0.$$

This implies that

$$\forall y \in \mathbb{C}^m, \forall x \in \text{Null}(A), \ x^* A^* y = 0.$$

Hence Null$(A) = (\text{Range}(A^*))^{\perp}$. ∎

Definition 2.13 *The rank of a matrix $A \in \mathbb{C}^{m \times n}$, $r = rank(A)$ is* dim$(Range(A))$, *i.e., the maximum number of linearly independent column vectors of the matrix A.*

Note that $r = \text{rank}(A)$ is also the maximum number of linearly independent row vectors of A, since from Proposition 2.5, it follows that

Proposition 2.6 dim$(Range(A)) = $ dim$(Range(A^*))$.

Proof. Given that $\mathbb{C}^n = \text{Null}(A) \oplus (\text{Null}(A))^{\perp} = \text{Null}(A) \oplus (\text{Range}(A^*))$, this implies that

$$\dim(\text{Range}(A)) = \dim(\text{Range}(A^*))$$

∎

2.3.1 Singular Values of a Matrix

Singular values of a matrix $A \in \mathbb{R}^{m \times n}$ (or $A \in \mathbb{C}^{m \times n}$) are related to the matrices AA^T and $A^T A$ (or AA^* and $A^* A$). For simplicity of presentation, we restrict our discussion to the case of real matrices.

Let $A \in \mathbb{R}^{m \times n}$ be a general rectangular matrix, then:

Proposition 2.7 *The matrices $AA^T \in \mathbb{R}^{m \times m}$ and $A^T A \in \mathbb{R}^{n \times n}$ are both symmetric positive semi-definite.*

Proof. Left for Exercise 2.12. ∎

Usually, $AA^T \neq A^T A$, even if $m = n$. In such case if $AA^T = A^T A$, the matrix A is said to be **normal**.

The matrices AA^T and $A^T A$ have the same positive eigenvalues as shown in what follows.

Proposition 2.8

$$\mathbb{R}_+ \cap \Lambda(AA^T) = \mathbb{R}_+ \cap \Lambda(A^T A).$$

Proof. Let σ be a positive eigenvalue for AA^T and let $x \neq 0$ be the corresponding eigenvector, i.e., $AA^T x = \sigma x$. Then $y = A^T x \neq 0$, since otherwise $AA^T x$ would be 0 and σ would not be a positive eigenvalue. Obviously, $A^T AA^T x = \sigma A^T x$ and thus, $A^T Ay = \sigma y$, i.e., $\{\sigma, y\}$ is a pair of eigenvalue, eigenvector for $A^T A$. ∎

Let now $r = \text{rank}(A) = \text{rank}(A^T)$. Then

Proposition 2.9 *The number of positive eigenvalues for AA^T and $A^T A$ is $r = rank(A) = rank(A^T)$.*

Proof. Note the following:

$$A : \mathbb{R}^n \to \mathbb{R}^m; \; A^T : \mathbb{R}^m \to \mathbb{R}^n,$$

and

$$A^T A : \mathbb{R}^n \to \mathbb{R}^n; \; AA^T : \mathbb{R}^m \to \mathbb{R}^m.$$

We first prove that $\text{Null}(A) = \text{Null}(A^T A)$:

If $x \in \text{Null}(A)$ then $x \in \text{Null}(A^T A)$. Conversely, suppose $x \in \text{Null}(A^T A)$, then $Ax \in \text{Range}(A)$ and $Ax \in \text{Null}(A^T) = (\text{Range}(A))^\perp$. Hence $Ax = 0$ and $x \in \text{Null}(A)$.

Since $A^T : \mathbb{R}^m \to \mathbb{R}^n$, then $\dim(\text{Range}(A^T)^\perp) + \dim(\text{Range}(A^T)) = n$. Thus, $\dim(\text{Range}(A^T))^\perp = n - r$. Since $\text{Null}(A) = (\text{Range}(A^T))^\perp$, and $\text{Null}(A) = \text{Null}(A^T A)$, then $\dim(\text{Null}(A^T A)) = n - r$. Hence $\dim(\text{rank}(A^T A)) = n - r$ and $\text{rank}(A^T A) = r$.

A similar argument applies to AA^T. ∎

2.3.2   Singular Value Decomposition

Based on the above, we can define *singular values* of any matrix A.

Definition 2.14 *Given a matrix $A \in \mathbb{R}^{m \times n}$ with $r = \text{rank}(A)$ and $p = \min\{m, n\}$, the set of p singular values of A, $\Sigma(A)$, is obtained by taking the square roots of the r positive eigenvalues of $\Lambda(AA^T)$ (or $\Lambda(A^T A)$) and completing these by $p - r$ zeros, i.e.,*

$$\Sigma(A) = \{\underbrace{\sigma_1 \, \sigma_2 \, ..., \sigma_r}_{r} \quad \underbrace{0...0}_{p-r}\}$$

with σ_i indexed in decreasing order:

$$\sigma_1 > \sigma_2 > ... > \sigma_r > 0, \text{ and } \sigma_{r+1} = \sigma_{r+2} = ... = \sigma_p = 0.$$

We state and prove now the singular value decomposition theorem, SVD, for any m by n matrix A that for simplicity, is restricted to have all its coefficients in $\mathbb{R}$. The complex coefficients version is left to Exercise 2.13.

Theorem 2.7 *Given $A \in \mathbb{R}^{m \times n}$ with $\text{rank}(A) = r$. There exist two orthogonal matrices $U \in \mathbb{R}^{m \times m}$ and $V \in \mathbb{R}^{n \times n}$, with $UU^T = I_m$ and $VV^T = I_n$ (I_m and I_n being unit matrices in $\mathbb{R}^m$ and $\mathbb{R}^n$, respectively) and a diagonal matrix $S \equiv S_{m,n} \in \mathbb{R}^{m \times n}$, such that*

$$AV = US,$$

or equivalently

$$A = USV^T \text{ or } S = V^T AU.$$

where $\Sigma(A) = \text{diag}(S)$

Using Jordan-Schur's theorem on $AA^T \in \mathbb{R}^{m \times m}$, one has:

$$AA^T U = UD, \tag{2.13}$$

where U and D are m by m matrices with U orthogonal ($UU^T = U^T U = I_m$) and D diagonal:

$$D = \begin{pmatrix} \{diag(\{\sigma_i^2 | i = 1, ..., r\}) & O_{r,m-r} \\ O_{m-r,r} & O_{r,r} \end{pmatrix}$$

where $O_{p,q}$ denotes a zero matrix with p rows and q columns and:

$$\sigma_1^2 \geq \sigma_2^2 \geq ... \geq \sigma_r^2.$$

Let $D^{1/2}$ be the square root of D with D_r and S_r be $r \times r$ the diagonal matrices extracted respectively from D and $D^{1/2}$. The diagonal entries of S_r are precisely:

$$\sigma_1 \geq \sigma_2 \geq ... \geq \sigma_r.$$

By writing $U = [U_r, U_{m-r}]$, where U_r are the first orthonormal columns of U and U_{m-r} are the last $m - r$ ones, then (2.13) is equivalent to:

$$AA^T U_r = U_r D_r \text{ and } AA^T U_{m-r} = O_{m,m-r}$$

Post-multiplying $AA^T U_r = U_r D_r$ by S_r^{-1} yields:

$$AA^T U_r S_r^{-1} = U_r S_r^2 S_r^{-1} = U_r S_r. \tag{2.14}$$

Letting $V_r = A^T U_r S_r^{-1}$, one has

$$V_r^T V_r = (A^T U_r S_r^{-1})^T (A^T U_r S_r^{-1}) = (S_r^{-1})^T U_r^T AA^T U_r S_r^{-1},$$

and since $AA^T U_r = U_r S_r^2$, thus,

$$V_r^T V_r = (S_r^{-1})^T U_r^T U_r S_r^2 S_r^{-1} = I_r.$$

Thus, from (2.14), one has

$$AV_r = U_r S_r. \tag{2.15}$$

Then, completing $V_r \in \mathbb{R}^{n \times r}$ into a $V \in \mathbb{R}^{n \times n}$ through a Gram-Schmidt procedure by adding $V_{r+1}, ..., V_n$, one has $AV_j = 0 \ \forall j > r$, since otherwise $\text{rank}(A) > r$. Thus, one obtains

$$V = [V_1, V_2, ..., V_n],$$

with

$$AV = \begin{pmatrix} U_r S_r & O_{m,n-r} \end{pmatrix}.$$

From the identity

$$\begin{pmatrix} U_r S_r & O_{m,n-r} \end{pmatrix} = \begin{pmatrix} U_r & U_{m-r} \end{pmatrix} \begin{pmatrix} S_r & O_{r,n-r} \\ O_{m-r,r} & O_{m-r,n-r} \end{pmatrix},$$

one has

$$AV = US,$$

where

$$S = \begin{pmatrix} S_r & O_{r,n-r} \\ O_{m-r,r} & O_{m-r,n-r} \end{pmatrix}.$$

∎

The singular value decomposition theorem has an "economy" version, restricted to the non-zero singular values and comes straight from (2.15) in the following form:

Theorem 2.8 *(Economy SVD) For any matrix $A \in \mathbb{R}^{m \times n}$, with $r = rank(A) \leq \min(m, n)$, there exist two matrices $U_r \in \mathbb{R}^{m \times r}$, $V_r \in \mathbb{R}^{n \times r}$ and $\Sigma_r \in \mathbb{R}^{r \times r}$ the diagonal matrix of the non-zero singular values of A, such that*

$$AV_r = U_r S_r,$$

with $U_r^T U_r = V_r^T V_r = I_r$, I_r being the r by r identity matrix.

Remark 2.2 *Note that* $U_r U_r^T \in \mathbb{R}^{m \times m}$ *and* $U_r U_r^T \neq I_m$, *with similarly* $V_r V_r^T \in \mathbb{R}^{n \times n}$ *and* $V_r V_r^T \neq I_n$.

The singular value decomposition is also implemented in MATLAB using the command svd

```
1. SVD     Singular value decomposition.
For a real A, [U,S,V] = svd(A) returns:
- A diagonal matrix S with same dimension as A
and diagonal elements <= 0 , in decreasing order,
- Two orthogonal matrices U and V such that
              A=U*S*V'.

2. S = svd(A) returns a vector containing
the singular values only.

3. [U,S,V] = svd(A,0) produces the "economy size" decomposition.
If A is m-by-n with m > n, then
- The first n columns of U are computed
   with S is n-by-n
```

The following is an example of the application of the singular value decomposition using the svd commands.

```
A =
      4       5       6
      7       8       9
      1       1      12
>> [U,S,V]=svd(A)
U =
   -0.4480    -0.2499    -0.8584
   -0.7031    -0.4945     0.5109
   -0.5522     0.8324     0.0458
S =
   19.1919          0          0
         0     6.9721          0
         0          0     0.2466
V =
   -0.3786    -0.5205     0.7653
   -0.4386    -0.6273    -0.6436
   -0.8151     0.5793    -0.0092
>> [Q,D]=eig(A*A')
Q =
    0.8584     0.2499     0.4480
   -0.5109     0.4945     0.7031
   -0.0458    -0.8324     0.5522
D =
```

```
      0.0608         0            0
           0    48.6105           0
           0         0     368.3287

>> [Q1,D1]=eig(A'*A)
Q1 =
      0.7653     0.5205      0.3786
     -0.6436     0.6273      0.4386
     -0.0092    -0.5793      0.8151
D1 =
      0.0608         0            0
           0    48.6105           0
           0         0     368.3287
>> Q=Q(:,[3 2 1])
Q =
      0.4480     0.2499      0.8584
      0.7031     0.4945     -0.5109
      0.5522    -0.8324     -0.0458
>> Q1=Q1(:,[3 2 1])
Q1 =
      0.3786     0.5205      0.7653
      0.4386     0.6273     -0.6436
      0.8151    -0.5793     -0.0092
>> D=D([3 2 1],[3 2 1])
D =
    368.3287         0            0
           0    48.6105           0
           0         0       0.0608
>> D1=D1([3 2 1],[3 2 1])
D1 =
    368.3287         0            0
           0    48.6105           0
           0         0       0.0608
```

2.4 Matrix Norms

Let $A \in \mathbb{K}^{m \times n} = \{a_{ij}, 1 \leq i \leq m, 1 \leq j \leq n\}$, where $\mathbb{K}$ is $\mathbb{R}$ or $\mathbb{C}$. A matrix norm is defined by the following properties:

Definition 2.15 *A matrix norm on* $\mathbb{K}^{m \times n}$ *is a function* $\|.\| : \mathbb{K}^{m \times n} \rightarrow [0, \infty)$, *satisfying the following properties:*

1. $||A|| \geq 0, \forall A \in \mathbb{K}^{m \times n}$, $(||A|| = 0 \Leftrightarrow A = 0)$,

2. $||A + B|| \leq ||A|| + ||B||, \forall A, B \in \mathbb{K}^{m \times n}$,

3. $||cA|| = |c|||A||, \forall c \in \mathbb{K}, \forall A \in \mathbb{K}^{m \times n}$.

4. $||A.B|| \leq ||A||.||B||, \forall A \in \mathbb{K}^{m \times n}, B \in \mathbb{K}^{n \times p}$.

There are 2 types of matrix norms:

1. **Matrix norms that are subordinate to vector norms** which satisfy all the 5 properties, specifically:

$$||A||_p = \max_{x \in \mathbb{R}^n} \frac{||Ax||_p}{||x||_p}.$$

In this context, one shows that

(a) for $p = 1$, $||A||_1 = \max_j \sum_i |a_{ij}|$. To prove this result, one starts with

$$y_i = \Sigma_{j=1}^n a_{ij} x_j.$$

Using the generalized triangle inequality, one concludes that

$$|y_i| \leq \Sigma_{j=1}^n |a_{ij}||x_j|,$$

and therefore,

$$\Sigma_{i=1}^m |y_i| \leq \Sigma_{i=1}^m \Sigma_{j=1}^n |a_{ij}||x_j| = \Sigma_{j=1}^n \Sigma_{i=1}^m |a_{ij}||x_j|.$$

Thus,

$$||y||_1 \leq \Sigma_{j=1}^n |x_j| \Sigma_{i=1}^m |a_{ij}| \leq \max_{1 \leq j \leq n} \Sigma_{i=1}^m |a_{ij}|||x||_1.$$

Hence for $\forall x \in \mathbb{R}^n ||x||_1 \neq 0$, one has

$$\frac{||Ax||_1}{||x||_1} \leq \max_{1 \leq j \leq n} \Sigma_{i=1}^m |a_{ij}|.$$

Therefore, by definition of $||A||_1$, one concludes that

$$||A||_1 \leq \max_{1 \leq j \leq n} \Sigma_{i=1}^m |a_{ij}|.$$

To prove that $||A||_1 = \max_{1 \leq j \leq n} \Sigma_{i=1}^m |a_{ij}|$, we seek a vector $\overline{x} \neq 0$ such that

$$||A\overline{x}||_1 = \max_{1 \leq j \leq n} \Sigma_{i=1}^m |a_{ij}|.$$

For that purpose, let j_0 be such that

$$\Sigma_{i=1}^m |a_{ij_0}| = \max_{1 \leq j \leq n} \Sigma_{i=1}^m |a_{ij}|.$$

Let $\bar{x}$ be such that $\bar{x}_i = 0$, $i \neq j_0$ and $\bar{x}_{j_0} = 1$. Let $\bar{y} = A\bar{x}$. Note that $\bar{y}_i = a_{ij_0}$, $\forall i$ and therefore,

$$||\bar{y}||_1 = \Sigma_{i=1}^m |a_{ij_0}| = \max_{1 \leq j \leq n} \Sigma_{i=1}^m |a_{ij}|.$$

∎

(b) For $p = \infty$, $||A||_\infty = \max_i \sum_j |a_{ij}|$. Proving this result is left as an exercise.

(c) For $p = 2$, $||A||_2 = \max svd(A)$ (maximum singular value of the matrix A).

One important result concerns the $||.||_2$ matrix norms of orthogonal and unitary matrices.

Proposition 2.10 *If $Q \in \mathbb{R}^{n \times n}$ is an orthogonal matrix, then $||Q||_2 = 1$. Similarly, if $U \in \mathbb{C}^{n \times n}$ is unitary then $||U||_2 = 1$.*

Proof. Left to Exercise 2.15. ∎

Subordinate matrix norms satisfy the inequality:

$$||Ax||_p \leq ||A||_p \times ||x||_p.$$

2. **Matrix norms that are vector-like norms**, the most used one being **Frobenius** norm:
$$||A||_F = \{\sum_{i,j} |a_{ij}|^2\}^{1/2},$$

which satisfy properties (1) to (4) and in particular,

$$||Ax||_2 \leq ||A||_F ||x||_2. \tag{2.16}$$

To prove such fact, consider the definition of $||Ax||_2$:

$$||Ax||_2^2 = \sum_{i=1}^m |\sum_{j=1}^n A_{ij}x_j|^2. \tag{2.17}$$

Using the Cauchy-Schwarz inequality, one shows that

$$|\sum_{j=1}^n A_{ij}x_j|^2 \leq \sum_{j=1}^n |A_{ij}|^2 \sum_{j=1}^n |x_j|^2 \leq \sum_{j=1}^n |A_{ij}|^2 ||x||_2^2,$$

which added to (2.17), leads to (2.16). ∎

Here are some MATLAB syntax for matrix norms:

norm(A) is the largest singular value of A, max(svd(A)).

norm(A,2) is the same as norm(A).

norm(A,1) is the 1-norm: the largest column max(sum(abs((A)))).

norm(A,inf), the infinity norm: the largest row sum max(sum(abs((A')))).

norm(A,'fro') is the Frobenius norm, sqrt(sum(diag(A'*A))).

norm(A,p) is available for matrix A only if p is 1, 2, inf or 'fro'.

2.5 Exercises

Exercise 2.1 *Find the eigenvalues and spectral radius of the following matrices:*

1. $A = [2\ 1; -1\ 1]$

2. $A = [2\ 1; 1\ 1]$

3. $A = [0\ 2\ 1; 1\ 2\ 1; -1\ 1\ 2]$

4. $A = [2\ 2\ 1; 1\ 2\ 1; 1\ 1\ 2]$

Exercise 2.2

Let u and v be 2 **non-zero** (column) vectors in $\mathbb{R}^n$. Let $A = I - uv^T$ where I is the n by n identity matrix.

1. Prove that the set of eigenvalues of A reduces to at most 2 numbers:

$$\Lambda(A) = \{\lambda_1, \lambda_2\},$$

where $\lambda_1 = 1$, $\lambda_2 = 1 - v^T u$. Which property should the vectors u and v satisfy in order for A to be singular (not invertible)?

2. Find the eigenspaces (the set of eigenvectors) E_1 and E_2 corresponding respectively to λ_1 and λ_2.

3. Show that when A is invertible, its inverse $A^{-1} = I + \alpha uv^T$. Find α.

Exercise 2.3

Let $v \in \mathbb{R}^n$, $v \neq 0$ and $a \in \mathbb{R}$, $a \neq 0$. Let:

$$A = I - a\,vv^T.$$

1. Show that A is symmetric.

2. Show that $\Lambda(A)$, the spectrum of A, consists of 2 distinct eigenvalues. Find the corresponding eigenvectors to each of these 2 eigenvalues.

3. Find the relation between a and $||v||$ that makes A spd.

Exercise 2.4

Let $F_k = [f_1, f_2, ..., f_k]$ a subset of orthonormal vectors in $\mathbb{R}^n$ ($F_k^T F_k = I_k$). Prove the identity:

$$I - F_k F_k^T = (I - f_k f_k^T)...(I - f_2 f_2^T)(I - f_1 f_1^T). \tag{2.18}$$

Exercise 2.5

Prove, for $U \in \mathbb{C}^{n \times n}$ unitary, that $|\det(U)| = 1$.

Exercise 2.6

Prove, for $Q \in \mathbb{R}^{n \times n}$ orthogonal, that $\det(Q) = \pm 1$.

Exercise 2.7

Prove, for $A \in \mathbb{C}^{n \times n}$ Hermitian, that $\Lambda(A) \subset \mathbb{R}$.

Exercise 2.8

Show, directly and through using Schur's decomposition, that for any n by n matrix, $p_A(\lambda)$ is a polynomial of degree exactly n.

Exercise 2.9

Show if $A \in \mathbb{R}^{n \times n}$ is spd, then $\Lambda(A) \subset \mathbb{R}_+$.

Exercise 2.10

Show if $A \in \mathbb{R}^{n \times n}$ is spsd, then $\Lambda(A) \subset \mathbb{R}_+ \cup \{0\}$.

Exercise 2.11

Prove that for $A \in \mathbb{R}^{n \times n}$:

$$A \text{ pd} \iff \frac{1}{2}(A + A^T) \text{ spd}$$

Exercise 2.12

Let $A \in \mathbb{R}^{m \times n}$ be a general rectangular matrix ($m < n$, $m = n$ or $m > n$). Prove that the matrices $AA^T \in \mathbb{R}^{m,m}$ and $A^T A \in \mathbb{R}^{n \times n}$ are both spsd with $\text{rank}(AA^T) = \text{rank}(A^T A) = \text{rank}(A)$.

Exercise 2.13

Reformulate Theorem 2.7 to have it apply for any complex matrix in $\mathbb{C}^{m \times n}$.

Exercise 2.14

Show that for $p = \infty$, $||A||_\infty = \max_i \sum_j |a_{ij}|$.

Exercise 2.15

Give the proof of Proposition 2.10.

Exercise 2.16

Show that the matrix Frobenius norm $||A||_F$ is not subordinate.

2.6   Computer Exercises

Computer Exercise 2.1 *Writing the classical Gram-Schmidt program to any set of n vectors in $\mathbb{R}^m$ or $\mathbb{C}^m$, $n \leq m$.*

Consider the MATLAB Algorithm 2.2 that takes for its input a set of n column vectors in the form of a non-zero $m \times n$ matrix V and outputs a set of $r \leq n$ orthonormal column vectors in the form of an $m \times r$ matrix F; a tolerance TOL is used to break the process whenever $r < n$ is reached and the set of column vectors of V are linearly dependent. The output parameter $k \leq n$ would give the order k at which the process would break (in such case $k < n$).

Algorithm 2.2 A Classical Gram-Schmidt Orthogonalization Process

---

```
function [F,r]=GramSchmidtClassical(V,tol)
% Input:
%       V an m by n matrix with rank r<=n<=m
%       a tolerance tol to reach r<n
%       All columns of V are non-zero
% Output:
%        r, rank of V
%        F an m by r matrix
%        IV an index vector that points
%        to first r linearly independent columns of V
[m,n]=size(V);F=zeros(m,n);r=n;IV=1:n;
for k=1:n % Induction process
   if k==1 % base step
       l=1;
       w=V(:,k);
   else % Project V(:,k) on span{F(:,1),...,F(:,l-1)}
       e=V(:,k);Fl=F(:,1:l-1);
       vc=(e'*Fl)'; % 2n(k-1) flops
       v=Fl*vc;%2n*(k-1)
       w=e-v;%n
   end
   a=norm(w);%2n flops
   if a>tol
       F(:,l)=(1/a)*w;%n flops
       IV(l)=k;l=l+1;
   else
       r=r-1;
   end
end
F=F(:,1:r);
```

---

1. Test the program on a set of matrices V obtained using the built-in matrices generators, `magic`, `rand`, `pascal`.
2. Test whether Algorithm 2.2 does (or does not) give orthonormal vectors $F(:, 1:r)$ for the cases run in 1.
3. Analyze the number of flops as a function of m and n.
4. Consider the case when the input matrices V are obtained by the function `rand(n)` for $n = 2^k$, $k = 5, ..., 10$. Check the execution time $T(n)$ using `tic toc` and the number of flops $f(n)$.
5. Extend Algorithm 2.2 to the case when the matrix V is complex, thus producing a unitary matrix F.
6. Modify this algorithm so as to use identity (2.18) obtaining therefore the `Modified Gram Schmidt` algorithm. Run this algorithm on the cases considered in 1 and test whether one obtains (or does not obtain) orthonormal vectors $F(:, 1:r)$.

Computer Exercise 2.2 *Algorithm 2.3 is a simulation of the proof of Theorem 2.4. It uses the built-in* MATLAB `eigs` *function and also the Classical Gram Schmidt Algorithm 2.2.*

Test Algorithm 2.3 for the cases of matrices generated by the built-in functions `rand`, `magic`, `pascal`.

Algorithm 2.3 Implementation of Schur's Decomposition

```
function   [F,T] = myschur(A)
% Input: a square n by n matrix
% Output: An upper triangular matrix T
%         A unitary matrix F
% The procedure uses the MATLAB function
% eigs to seek the largest eigenvalue of a square matrix
n = length(A);d=zeros(n,1);% d storeS the eigenvalues of A
T=zeros(n);k=1;
while k<=n-1
    [v,lambda] = eigs(A,1,'lm');d(k)=lambda;
    E = [v eye(n-k+1)];F1 = GramSchmidtClassical(E,0.5*10^(-5));
    A1 = F1'*A*F1;T(k:n,k:n)=A1;
    % Updating the matrix F
    if k==1
       F=F1;
    else
       F(:,k:n)=F(:,k:n)*F1;T(1:k-1,k:n)=T(1:k-1,k:n)*F1;
    end
    % Update matrix A
    A=A1(2:n-k+1,2:n-k+1);k=k+1;
end
```

Computer Exercise 2.3 *In the following commands:*

```
[U,S,V]=svd(A);
[Q,D]=eig(A*A');
[Q1,D1]=eig(A'*A);
```

find (if they exist) the relationships between the matrices Q, Q1, D, D1, U, S, V. Justify your answer.

Computer Exercise 2.4

1. Implement the proof of Theorem 2.7 by completing MATLAB Algorithm 2.4, which takes as input a rectangular matrix A and outputs two orthogonal matrices $U \in \mathbb{R}^{m \times m}$, and $V \in \mathbb{R}^{n \times n}$ and a diagonal matrix $S \in \mathbb{R}^{m \times n}$ consisting of the singular values of A, such that $AV = US$. The procedure uses MATLAB schur(X,'real') and the already written Algorithm 2.2.

2. Test Algorithm 2.4 by generating a rectangular matrix using the function and compare the results with those obtained using MATLAB svd.

Algorithm 2.4 Singular Value Decomposition Using MATLAB Commands

```
function [U S V] = mysvd(A)
[m n] = size(A);
[U T] = schur(A*A','real');%AA'U=U*T
S = zeros(m,n);
r=rank(T);% r<=m
s=sqrt(diag(T));
s=sort(s,'descend');
%%%%%%%%%%%%%% %%%%



            COMPLETE THIS PART OF THE PROGRAM



%%%%%%%%%%%%%% %%%%
V=GramSchmidtClassical(V,0.5*10^(-6));
```

Chapter 3

Gauss Elimination and LU Decompositions of Matrices

In this chapter, our main objective is to prove the general **decomposition of any matrix into the product of a unit lower triangular matrix by an upper triangular one**. Such decompositions formalize the procedures for solving systems of linear equations by Gauss elimination. This material is at the heart of matrix computations and constitutes the basis for computing determinant and inverse of square matrices. More specific details can be found in books such as Ciarlet [19], Poole [56], Trefethen [69], and Golub and Van Loan [33].

Gauss reduction is the main procedure for obtaining an LU decomposition or PLU decompositions. In the first case, the reduction is said to be a *Naive Gauss Reduction* (Section 3.3) while in the second case (Section 3.4), the reduction is referred to as *Gauss Reduction with Partial Pivoting*, scaled or unscaled.

Unless stated otherwise, all matrices are considered to be **dense, real and square matrices**.

3.1 Special Matrices for LU Decomposition

We introduce first two types of matrices that play a key role in this chapter, specifically triangular[1] and permutation matrices.

3.1.1 Triangular Matrices

Definition 3.1 *An $m \times n$ matrix A is said to be lower (upper) triangular if $A_{ij} = 0, i < j, (i > j)$.*

---

[1] Upper triangular matrices were already obtained in the Gram-Schmidt process (Chapter 2, Theorem 2.2.)

The following properties for triangular matrices are stated for lower type, but they are also valid for upper triangular matrices.

Theorem 3.1 *Let L, $M \in \mathbb{R}^{n \times n}$ be lower triangular matrices, then:*

1. *$L + M$ is lower triangular.*

2. *$N = L \times M$ is lower triangular. Furthermore, if $L_{ii} = M_{ii} = 1$, $\forall i$, then $N_{ii} = 1, \forall i$.*

3. *If L is a lower triangular invertible matrix, then L^{-1} is also lower triangular. Furthermore, if $L_{ii} = 1, \forall i$, then $L_{ii}^{-1} = 1, \forall i$.*

Proof. Left to be solved in Exercise 3.3.   ■

Note that Property 1 of this Theorem is also valid when L and M are triangular and non-square, i.e., rectangular matrices.

3.1.2   Permutation Matrices

Definition 3.2 *Let $I \in \mathbb{R}^{n \times n}$ be the identity matrix,*

$$I = \begin{pmatrix} 1 & 0 & 0 & 0 & \ldots & 0 \\ 0 & 1 & . & 0 & \ldots & 0 \\ . & \ldots & . & . & \ldots & . \\ 0 & \ldots & 0 & 0 & 1 & 0 \\ 0 & \ldots & \ldots & 0 & 0 & 1 \end{pmatrix} = [e_1, e_2, \ldots, e_n] = \begin{pmatrix} e_1^T \\ e_2^T \\ \ldots \\ \ldots \\ e_n^T \end{pmatrix} \quad (3.1)$$

I is identified with the set of naturally ordered first n integers $\mathcal{I}_n = \{1, 2, \ldots, n\}$. Let then $\mathcal{P}_n = \{i_1, i_2, \ldots, i_n\}$ be any permutation set $\mathcal{I}_n$. Then:

1. *The column-permutation matrix $Q(i_1, i_2, \ldots, i_n)$ is the $n \times n$ matrix obtained from (3.1), through the permutation of the columns of I according to $\mathcal{P}_n = \{i_1, i_2, \ldots, i_n\}$, specifically:*

$$Q = [e_{i_1} \; e_{i_2} \ldots e_{i_n}]$$

2. *The row-permutation matrix $P(i_1, i_2, \ldots, i_n)$ is the $n \times n$ matrix obtained from (3.1), through the permutation of the rows of I according to $\{i_1, i_2, \ldots, i_n\}$, specifically:*

$$P = \begin{pmatrix} e_{i_1}^T \\ e_{i_2}^T \\ \ldots \\ \ldots \\ e_{i_n}^T \end{pmatrix}$$

One proves the following:

Proposition 3.1 *Given any permutation* $\mathcal{P}_n = \{i_1, i_2, ..., i_n\}$ *of* $\mathcal{I}_n = \{1, 2, ..., n\}$, *let* Q *and* P *be respectively the columns and row permutation matrices associated with* $\mathcal{P}_n$. *Then:*

1. $Q = P^T$ *and* $Q \times P = P \times Q = I$, *i.e.,* $P^{-1} = Q$.

2. *If* $A \in \mathbb{R}^{n \times n}$, *then* $P \times A$ *results in the permutation of the rows of* A *according to* $\{i_1, i_2, ..., i_n\}$.

3. $A \times Q$ *results in the permutation of the columns of* A *according to* $\{i_1, i_2, ..., i_n\}$.

Proof. The first part of the proof is left to Exercise 3.1. For the second part, consider the lth row of $P \times A$. Its kth element is given by $e_{i_l}^T A(:, k) = A(i_l, k)$. Hence, the elements on the lth row of PA are precisely those of row i_l in A. A similar argument applies for $A \times Q$, by looking to the mth column of AQ and finding that it is exactly the i_m^{th} column of A. ∎

General permutation matrices are obtained from the product of elementary permutation matrices which result from the permutation of only 2 indices in $\mathcal{I}_n = \{1, 2, ..., n\}$.

Definition 3.3 *An elementary permutation matrix* $P(i, j)$ *is obtained from the identity matrix* I *through the permutation of rows (or columns)* i *and* j.

Note that $P(i, j)$ is symmetric and equal to its inverse. Any given permutation $\mathcal{P}_n = \{i_1, i_2, ..., i_n\}$ of $\mathcal{I}_n$ can be obtained through a sequence of $n - 1$ elementary permutations of the form:

$$1 \rightarrow i_1$$
$$2 \rightarrow i_2$$
$$....$$
$$...$$
$$n - 1 \rightarrow i_{n-1}$$

As an example, let $n = 4$. Consider the permutation $\{3, 4, 2, 1\}$ obtained on $\{1, 2, 3, 4\}$. It results from a sequence of 3 elementary permutations:

Index Changes	Associated Permutation	Elementary Matrix
$1 \rightarrow 3$	$\{1, 2, 3, 4\} \rightarrow \{3, 2, 1, 4\}$	$P(1, 3)$
$2 \rightarrow 4$	$\{3, 2, 1, 4\} \rightarrow \{3, 4, 1, 2\}$	$P(2, 4)$
$3 \rightarrow 4$	$\{3, 4, 1, 2\} \rightarrow \{3, 4, 2, 1\}$	$P(3, 4)$

Thus, the matrix $P \equiv P(\{3, 4, 2, 1\})$ is given by:

$$P \equiv P(\{3, 4, 2, 1\}) = P(3, 4)P(2, 4)P(1, 3).$$

Hence given a matrix $A \in \mathbb{R}^{n \times n}$, $P(3, 4)P(2, 4)P(1, 3)A$ would lead to the successive row permutations:

1. Rows 1 and 3 of A,

2. Rows 2 and 4 of $P(1,3)A$,

3. Rows 3 and 4 of $P(2,4)P(1,3)A$.

As a result, the rows of the matrix A are ordered according to the permutation $\{3\ 4\ 2\ 1\}$. As an example, starting in MATLAB with the matrix A:

```
0.9355    0.0579    0.1389    0.2722
0.9169    0.3529    0.2028    0.1988
0.4103    0.8132    0.1987    0.0153
0.8936    0.0099    0.6038    0.7468
```

then

$$P(3,4)P(2,4)P(1,3)A,$$

give:

```
0.4103    0.8132    0.1987    0.0153
0.8936    0.0099    0.6038    0.7468
0.9169    0.3529    0.2028    0.1988
0.9355    0.0579    0.1389    0.2722
```

Note also that these row permutations can be simply completed using the MATLAB commands:

```
I=[3 4 2 1];
A(1:4,:)=A(I,:);
```

3.2 Gauss Transforms

3.2.1 Preliminaries for Gauss Transforms

We start by considering the matrix obtained by modifying the identity matrix $I \in \mathbb{R}^{n \times n}$ by a rank one external product matrix:

$$M = M(u,v) = I - uv^T \tag{3.2}$$

where u, v are non-zero vectors in $\mathbb{R}^n$ (see Exercise 2.2). Let x be also a non-zero vector in $\mathbb{R}^n$. Then:

$$y = Mx = x - (v^T x)u.$$

Assume there exists $1 \le k < n$ with:

$$(x_{k+1}, ..., x_n)^T \ne 0. \tag{3.3}$$

We intend to choose u and v in (3.2), such that

$$y_i = x_i,\ 1 \le i \le k \text{ and } y_i = 0,\ i = k+1, ..., n, \tag{3.4}$$

i.e., the matrix $M = M^{(k)}$ "annihilates" all the components of x with index starting at $k+1$.
The first part of (3.4) is satisfied by choosing u, such that

$$u_i = 0,\ i = 1, ..., k. \tag{3.5}$$

On the other hand, to obtain the second part necessitates that

$$x_i = (v^T x)u_i,\ i = k+1, ..., n. \tag{3.6}$$

This leads to the following result.

Proposition 3.2 *Let x and v be non-zero vectors in $\mathbb{R}^n$, such that x satisfies (3.3) and $v^T x \ne 0$, (i.e., v not orthogonal to span$\{x\}$). Let then:*

$$u^{(k)} = \frac{1}{v^T x} \begin{pmatrix} 0 \\ \vdots \\ 0 \\ x_{k+1} \\ \vdots \\ x_n \end{pmatrix},$$

then $M^{(k)} = M^{(k)}(u^{(k)}, v) = I - u^{(k)}v^T$ satisfies:

$$M^{(k)}x = \begin{pmatrix} x_1 \\ \vdots \\ x_k \\ 0 \\ \vdots \\ 0 \end{pmatrix}.$$

Proof. It follows from (3.3), (3.4), (3.5) and (3.6). ∎

3.2.2 Definition of Gauss Transforms

In addition to condition (3.3) on x, let us further assume that

$$x_k \ne 0. \tag{3.7}$$

By choosing $v = v^{(k)} = e_k \in \mathbb{R}^n$ (the unit vector that has all its components 0, except the kth element equal to 1), one has $(v^{(k)})^T x = e_k^T x = x_k \ne 0$ and therefore in applying Proposition 3.2, one obtains from (3.6):

$$u_i = \frac{x_i}{x_k},\ i = k+1, ..., n.$$

Consequently, let $u^{(k)} = \alpha^{(k)} \in \mathbb{R}^n$, such that

$$(\alpha^{(k)})^T = (0, ..., 0, \alpha^{(k)}_{k+1}, ..., \alpha^{(k)}_n), \text{ where } \alpha^{(k)}_i = \frac{x_i}{x_k}, i = k+1, ..., n.$$

A Gauss transform can then be defined as follows:

Definition 3.4 *Let $M^{(k)} \in \mathbb{R}^{n \times n}$ be such that*

$$M^{(k)} = I - \alpha^{(k)} e_k^T,$$

$\alpha^{(k)}$ *is said to be the Gauss vector with its components $\alpha^{(k)}_i, i = k+1, ..., n$ being the "multiplying factors" of the kth level Gauss transform.*

Note that $M^{(k)}$ is a lower triangular matrix of the form:

$$M^{(k)} = \begin{pmatrix} 1 & ... & 0 & 0 & ... & 0 \\ . & ... & . & . & ... & . \\ 0 & ... & 1 & 0 & ... & 0 \\ 0 & ... & -\alpha^{(k)}_{k+1} & 1 & ... & 0 \\ . & ... & . & . & ... & . \\ 0 & ... & -\alpha^{(k)}_n & 0 & ... & 1 \end{pmatrix}$$

Thus, from Proposition 3.2 one has:

$$M^{(k)} x = \begin{pmatrix} x_1 \\ . \\ . \\ x_k \\ 0 \\ . \\ 0 \end{pmatrix},$$

with furthermore

$$(M^k)^{-1} = I + \alpha^{(k)} e_k^T.$$

This identity is verified using $e_k^T \alpha^{(k)} = 0$ in

$$(I - \alpha^{(k)} e_k^T)(I + \alpha^{(k)} e_k^T) = I - (\alpha^{(k)} e_k^T)(\alpha^{(k)} e_k^T) = I - \alpha^{(k)}(e_k^T \alpha^{(k)}) e_k^T = I.$$

3.3 Naive *LU* Decomposition for a Square Matrix with Principal Minor Property (pmp)

To solve the system of linear equations $Ax = b$, Gauss transforms allow to reduce the solving procedure into that of two triangular systems the first of which being lower triangular and the second upper triangular. Without loss in generality, we shall consider first the case of square matrices having the "principal minor property" (pmp).

Definition 3.5 *A square $n \times n$ matrix A has the principal minor property (pmp) if each of the principal sub-matrices $A_{ii}, i = 1, ...n$ is also invertible.*

Assume $A \in \mathbb{R}^{n \times n}$ has the pmp. We show in Theorem 3.2 that A can be decomposed as the product $L \times U$ where L is lower triangular and U upper triangular, with $L_{ii} = 1, \forall i$. To obtain the product LU, we start by seeking a sequence of transforms, called Gauss transforms, $\{M^{(k)} \mid k = 1, ..., (n-1)\}$ such that the product $\prod_{i=1}^{n-1} M^{(i)} A = U$, where U is an upper triangular matrix.

Consider now the successive application of a sequence of Gauss transforms on a matrix $A \in \mathbb{R}^{n \times n}$ having the principal minor property (pmp). For that purpose, let A be written in terms of its column vectors:

$$A = (A(:, 1)\, A(:, 2)\, ...A(:, n-1)\, A(:, n)),$$

where $A(:, j) \in \mathbb{R}^n$ is the jth column vector of the matrix A. As a first step, consider the following transform:

$$M^{(1)} = I - \alpha^{(1)} e_1^T, \alpha^{(1)} = (0, \alpha_2, ..., \alpha_n)^T,$$

$$M^{(1)} = \begin{pmatrix} 1 & 0 & . & . & 0 \\ -\alpha_2 & 1 & 0 & . & 0 \\ . & 0 & 1 & . & 0 \\ . & . & . & . & . \\ -\alpha_n & 0 & . & 0 & 1 \end{pmatrix}$$

Consequently, $M^{(1)}$ is a modification of the unit matrix by $-\alpha^{(1)} e_1^T$, which is a rank 1 matrix.
In the case when $\alpha^{(1)} = (0, a_{21}/a_{11}, ..., a_{n1}/a_{11})^T$, one obtains:

$$M^{(1)} A = A - \alpha^{(1)} e_1^T A = \begin{pmatrix} a_{11} & a_{12} & . & . & a_{1n} \\ 0 & a_{22}^{(1)} & . & . & a_{2n}^{(1)} \\ . & & . & . & . \\ 0 & a_{n2}^{(1)} & . & . & a_{nn}^{(1)} \end{pmatrix}$$

The equivalent "column" writing of $M^{(1)} A$ is given by:

$$(M^{(1)} A)(:, j) = ((I - \alpha^{(1)} e_1^T) A)(:, j) = A(:, j) - \alpha^{(1)} e_1^T A(:, j) = A(:, j) - A(1, j) \alpha^{(1)}$$

Given that $a_{i1}^{(1)} = 0$, this is equivalent to writing:

$$a_{ij}^{(1)} = \begin{cases} a_{ij}, & \text{if } i = 1 \\ a_{ij} - \alpha_i^{(1)} a_{1j}, & \text{if } 2 \le i, j \le n \end{cases}$$

We thus obtain $A^{(1)}$ from $M^{(1)} A \in \mathbb{R}^{n \times n}$, such that

$$
A^{(1)} = \begin{pmatrix}
a_{11} & a_{12} & \cdot & \cdot & a_{1n} \\
0 & a_{22}^{(1)} & \cdot & \cdot & a_{2n}^{(1)} \\
0 & \cdot & \cdot & \cdot & \cdot \\
\cdot & \cdot & \cdot & \cdot & \cdot \\
0 & a_{n2}^{(1)} & \cdot & \cdot & a_{nn}^{(1)}
\end{pmatrix}
$$

To obtain $M^{(2)}$, the same procedure can be repeated provided that $a_{22}^{(1)} \neq 0$. This is guaranteed by the principal minor property (pmp) and is left to Exercise 3.4. Thus, recursively, we obtain a sequence of Gauss transforms: $\{M^{(1)}, M^{(2)}, ..., M^{(n-1)}\}$ where $M^{(l)} = (I - \alpha^{(l)} e_l^T)$, $l = 1, ..., n-1$.

Remark 3.1 *At the kth step of the process and for $k = 1, ..., n-1$, we reach the following:*

1. *A matrix $A^{(k-1)} = M^{(k-1)} ... M^{(1)} A$ that is upper triangular from the 1st to the kth column.*

2. *In the matrix $A^{(k)}$, it is necessary to have $a_{kk}^{(k-1)} \neq 0$, in order to continue the triangulation process.*

This leads to the following:

Lemma 3.1 *Let $A \in \mathbb{R}^{n \times n}$ with the principal minor property (pmp). There exists a sequence of Gauss transforms $\{M^{(k)} \mid k = 1, ..., (n-1)\}$, such that*

$$
M^{(n-1)} ... M^{(2)} M^{(1)} A = U \tag{3.8}
$$

is an upper triangular matrix.

Furthermore $[M^{(k)}]^{-1} = I + \alpha^{(k)} e_k^T$ is also a lower triangular matrix. Thus, using also Equation (3.8), one obtains the following identity:

$$
A = (M^{(1)})^{-1} (M^{(2)})^{-1} ... (M^{(n-1)})^{-1} U.
$$

which is equivalent to:

$$
A = (I + \alpha^{(1)} e_1^T)(I + \alpha^{(2)} e_2^T) ... (I + \alpha^{(n-1)} e_{n-1}^T) U \tag{3.9}
$$

The following lemma is obtained by induction.

Lemma 3.2 *Let $L = \prod_{k=1}^{n-1} (I + \alpha^{(k)} e_k^T)$. Then:*

$$
L = I + \sum_{k=1}^{n-1} \alpha^{(k)} e_k^T
$$

The previous lemmas concur to give the following factorization theorem.

Theorem 3.2 LU Decomposition for a pmp Square Matrix. *Let $A \in \mathbb{R}^{n \times n}$ possess the principal minor property (pmp). Then there exists a lower triangular matrix L, with diagonal elements all equal to 1 and an upper triangular matrix U, both in $\mathbb{R}^{n \times n}$ such that*

$$A = LU.$$

Furthermore, such decomposition is unique.

Proof. The unicity of the decomposition is proven by contradiction. Assume the existence of L_i, U_i, $i = 1, 2$ such that $A = L_1 U_1 = L_2 U_2$, with the diagonal elements of L_i, $i = 1, 2$, all equal to 1 and thus invertible. Since A has the pmp, U_I $I = 1, 2$ are invertible also. Hence:

$$(L_2)^{-1} L_1 = U_2 (U_1)^{-1}. \tag{3.10}$$

In this last identity, note that the left-hand side is a product of 2 lower triangular matrices and is thus also a lower triangular one. Similarly the right-hand side is a product between 2 upper triangular matrices, resulting also in an upper triangular matrix. Hence:

$$(L_2)^{-1} L_1 = U_2 (U_1)^{-1} = D,$$

where D is a diagonal matrix. Furthermore, one easily shows (Exercise 3.5) that the elements of the product $(L_2)^{-1} L_1$ are all equal to 1. Hence $D = I$ and therefore:

$$(L_2)^{-1} L_1 = U_2 (U_1)^{-1} = I,$$

leading to: $L_1 = L_2$ and $U_1 = U_2$. ∎

3.3.1 Algorithm and Operations Count

Implementation of Naive Gauss Reduction leads to Algorithm 3.1.

Algorithm 3.1 *LU* Decomposition for A with Principal Minor Property
---
```
function [L,U]=mylu1(A)
%A has the principal minor property
n=size(A);
for k=1:n-1
        piv=1/A(k,k);
        A(k+1:n,k)=piv*A(k+1:n,k);
        A(k+1:n,k+1:n)=A(k+1:n,k+1:n)-A(k+1:n,k)*A(k,k+1:n);
end
L=eye(n)+tril(A,-1);U=triu(A);
```

Operations Count

Each application of a Gauss Transform requires the following:

- one division for the pivot,
- $n - k$ multiplications for the multipliers,
- $(n - k)^2$ additions and $(n - k)^2$ multiplications for:
$$A(k+1:n,k+1:n)-A(k+1:n,k)*A(k,k+1:n).$$

Consequently, the resulting total number of arithmetic operations is given by:

$$\Sigma_{k=1}^{n-1}1 + (n - k) + 2(n - k)^2 = O(\frac{2n^3}{3}). \qquad (3.11)$$

The proof is left to Exercise 3.7.

Storage requirement is precisely $O(n^2)$, as the multipliers are stored in the lower part of the matrix A.

3.3.2 LDL_1^T Decomposition of a Matrix Having the Principal Minor Property (pmp)

We consider a square matrix A having the pmp. Given the previous decomposition theorem, A can be written as: $A = L \times U$. Since:

$$\det(A) = \det(L) \times \det(U),$$

then $\det(U) \neq 0$, and $u_{ii} \neq 0$, $\forall i$. Define (using MATLAB notations) the diagonal matrix:

$$D = \texttt{diag(diag}(U)),$$

where $\texttt{diag(diag}(U))$ extracts the diagonal part of the matrix A. Obviously $A = LDD^{-1}U$. Setting $L_1 = (D^{-1}U)^T = U^T D^{-1}$, a **variant** of the LU decomposition theorem would be an "LDL_1^T" version, summarized in:

Theorem 3.3 *Assume a square matrix $A \in \mathbb{R}^{n \times n}$ has the principal minor property (pmp). Then there exists a diagonal matrix D, and two lower triangular matrices L and L_1, which diagonal elements are all 1, such that*

$$A = LDL_1^T.$$

Furthermore, $\det(A) = \det(D)$.

3.3.3   The Case of Symmetric and Positive Definite Matrices: Cholesky Decomposition

We consider successively the cases of **symmetric** and then **symmetric positive definite matrices**.

1. **Symmetric Matrices having pmp.** In such case, one has $A = A^T$. The operations can be reduced by half. See Exercises 3.8, 3.9 and Computer Exercise 3.1.

2. **Symmetric Positive Definite**

Definition 3.6 *A matrix $A \in \mathbb{R}^{n \times n}$ is said to be* **positive definite** *if:*

$$x^T A x > 0, \forall x \in \mathbb{R}^n, \ x \neq 0, \ (x^T A x = 0, \ if \ and \ only \ if \ x = 0).$$

An immediate consequence of this definition is the following proposition.

Proposition 3.3 *A positive definite matrix has the principal minor property (pmp).*

Proof. Select $x \in \mathbb{R}^{n \times n}$, such that $x \neq 0$ and $x_k = 0$, $k > i$, $1 < i \leq n$. Then:

$$0 < x^T A x = (x^{(i)})^T A_i x^{(i)}$$

where A_i is the ith principal minor of the matrix A and $x^{(i)} \in \mathbb{R}^i$, $x_k(i) = x_k$, $k \leq i$.   ∎

As such the LDL^T decomposition is valid for symmetric and positive definite matrices. Furthermore, the positive definiteness of A implies the positive definiteness of D. Since D is a diagonal matrix, all its elements d_i, $1 \leq i \leq n$, would be necessarily positive. Let D_1 be the diagonal matrix which elements are $\sqrt{d_i}$, $1 \leq i \leq n$. Obviously:

$$D_1 \times D_1 = D,$$

and the LDL^T decomposition of A is equivalent to

$$A = L D_1 D_1 L^T = G G^T,$$

with $G = LD_1$. This proves the **Cholesky decomposition theorem**:

Theorem 3.4 *A symmetric and positive definite matrix A admits the decomposition $A = GG^T$, where the matrix G is lower triangular. Furthermore, such decomposition is unique.*

The proof of the uniqueness of Cholesky's decomposition is left to Exercise 3.6.   ∎

Obtaining algorithms for Cholesky's decomposition is left to Computer Exercise 3.3.

3.3.4   Diagonally Dominant Matrices

Definition 3.7 *A square matrix A is said to be*

1. **Diagonally dominant** *if:* $|a_{ii}| \geq \Sigma_{j=1, j\neq i}^{n}|a_{ij}|, \forall i$.

2. **Strictly diagonally dominant** *if:* $|a_{ii}| > \Sigma_{j=1, j\neq i}^{n}|a_{ij}|, \forall i$.

Being diagonally dominant with positive diagonal elements implies the principal minor property (pmp) for matrices.

Proposition 3.4 *If a matrix $A \in \mathbb{R}^{n\times n}$ is a square symmetric matrix with* **positive diagonal elements**, *then:*

1. *Strict diagonal dominance for A implies that A is symmetric positive definite (spd).*

2. *Diagonal dominance implies that A is symmetric positive semi-definite (spsd).*

Proof. The proof is based on the main fact that a symmetric matrix A is positive definite (semi-definite) if and only if $\sigma(A) \subset (0, \infty)$ $(\sigma(A) \subset [0, \infty))$. Note that if A is strictly diagonally dominant, then if $\lambda \in \sigma(A)$, then there exists $x \neq 0$, such that $Ax = \lambda x$. If $|x_{i_0}| = \max_i\{|x_i|\}$, then for such i_0:

$$(a_{i_0 i_0} - \lambda)x_{i_0} = -\Sigma_{j\neq i_0}a_{i_0 j}x_j.$$

Hence:

$$|(a_{i_0 i_0} - \lambda)||x_{i_0}| \leq \Sigma_{j\neq i_0}|a_{i_0 j}||x_j| \leq |x_{i_0}|\Sigma_{j\neq i_0}|a_{i_0 j}|.$$

Thus:

$$|(a_{i_0 i_0} - \lambda)| \leq \Sigma_{j\neq i_0}|a_{i_0 j}|,$$

and therefore,

$$a_{i_0 i_0} - \Sigma_{j\neq i_0}|a_{i_0 j}| \leq \lambda \leq a_{i_0 i_0} + \Sigma_{j\neq i_0}|a_{i_0 j}|,$$

which proves the positive definiteness of λ when one has strict dominance and positive semi-definiteness otherwise. ∎

Remark 3.2 *The above result does not cover the standard tridiagonal matrix:*

$$T = \begin{pmatrix} 2 & -1 & 0 & ... & 0 & 0 \\ -1 & 2 & -1 & 0 & .. & 0 \\ . & . & . & . & . & . \\ . & . & . & . & . & . \\ 0 & . & 0 & -1 & 2 & -1 \\ 0 & . & . & 0 & -1 & 2 \end{pmatrix} \quad (3.12)$$

This matrix is diagonally dominant only. However, we can show its positive-definiteness by computing directly $x^T T x$. One finds that

$$x^T T x = x_1^2 + \Sigma_{i=2}^{n-1}(x_i - x_{i-1})^2 + x_n^2.$$

and $x^T T x \geq 0$ with $x^T T x = 0$ if and only if $x = 0$.

3.4 *PLU* Decompositions with Partial Pivoting Strategy

The general case of a square matrix A (invertible or not invertible) requires introducing at each reduction k, $k = 1 : n - 1$, an elementary permutation matrix $P^{(k)} \equiv P(k, j)$, $j \geq k$, such that $\prod_{k=1}^{n-1} M^{(k)} P^{(k)} A = L \times U$. Each permutation matrix $P^{(k)}$ is obtained on the basis of a "pivoting" strategy that selects $j \geq k$ and then perform the permutation of rows j and k of a matrix through its pre-multiplication by $P(k, j)$.

The technique of "partial pivoting" is used to handle square matrices of any type, particularly those with no (pmp) property, for example when $a_{11} = 0$, or in general when $a_{11} \neq 0$ in view of reducing the effect of rounding errors on the computations. In the sequel, two types of strategies are explained: unscaled and scaled partial pivoting.

3.4.1 Unscaled Partial Pivoting

Let $p_1 = \mid a_{i_1 1} \mid = \max_{1 \leq i \leq n} \mid a_{i1} \mid$.
Then, on the basis of a user's tolerance ϵ_{tol} selected to detect the possible "singularity" of the matrix A, two situations may occur:

1. $p_1 > \epsilon_{tol}$. In that case, we define the elementary permutation matrix $P^{(1)} = P(1, i_1)$, ($P^{(1)} = I$ if $i_1 = 1$) and consider:

$$B^{(1)} = P^{(1)} A = \begin{pmatrix} a_{i_1 1} & . & . & a_{i_1 n} \\ a_{21} & . & . & a_{2n} \\ . & . & . & . \\ a_{11}. & . & . & a_{1n}. \\ . & . & . & . \\ a_{n1} & . & . & a_{nn} \end{pmatrix}$$

Let $M^{(1)}$ be the Gauss matrix associated with the vector $B^{(1)}(1 : n, 1)$, i.e., $M^{(1)} = I - \alpha^{(1)} e_1^T$, with

$$\alpha^{(1)} = \begin{pmatrix} 0 \\ B^{(1)}(2 : n, 1)/B^{(1)}(1, 1) \end{pmatrix}$$

2. If $p_1 \leq \epsilon_{tol}$, i.e., $p_1 \approx 0$, one sets $M^{(1)} = P^{(1)} = I$.

We thus obtain:

$$M^{(1)} P^{(1)} A = \begin{pmatrix} X & X & . & X \\ 0 & X & . & X \\ . & . & . & . \\ 0 & X & . & X \\ 0 & X & . & X \end{pmatrix},$$

and may proceed to the next columns. Hence, through a recurrence argument, two sequences of matrices may be constructed:

1. Permutation matrices $\{P^{(k)}\,|\,k = 1 : n - 1\}$,

2. Gauss matrices $\{M^{(k)}\,|\,k = 1 : n - 1\}$,

satisfying the equation:

$$M^{(n-1)}P^{(n-1)}M^{(n-2)}P^{(n-2)}...M^{(1)}P^{(1)}A = U, \qquad (3.13)$$

where U is an upper triangular matrix. Such technique of Gauss elimination with partial pivoting leads to the following theorem:

Theorem 3.5 *Let $A \in \mathbb{R}^{n \times n}$. There exist three matrices $\in \mathbb{R}^{n \times n}$: U upper triangular, L lower triangular with $L_{ii} = 1$ and P a permutation matrix verifying the following identity:*

$$PA = LU \ with \ P = P^{(n-1)}...P^{(1)}. \qquad (3.14)$$

Furthermore, if $\alpha^{(k)}$ is the set of multiplying factors at the kth elimination, the coefficients of the kth column of L (below the diagonal), are obtained from $L(k + 1 : n, k) = P^{(n-1)}...P^{(k+1)}\alpha^{(k)}$.

Proof. We illustrate it in the case when $n = 3$. Equation (3.13) becomes:

$$M^{(2)}P^{(2)}M^{1}P^{(1)}A = U, \qquad (3.15)$$

which is equivalent to:

$$M^{(2)}(I - P^{(2)}\alpha^{(1)}(P^{(2)}e_1)^T)P^{(2)}P^{(1)}A = U,$$

which is also equivalent to (since $P^{(2)}e_1 = e_1$):

$$M^{(2)}(I - P^{(2)}\alpha^{(1)}e_1^T)P^{(2)}P^{(1)}A = U.$$

Hence, if the original multipliers vectors of the Gauss transforms are $\alpha^{(k)}$, $k = 1 : n - 1$, then the permutations matrices P^j, $j = 1 : k - 1$ transform the $\alpha^{(k)}$ as follows:

$$\tilde{\alpha}^{(k)} = P^{(n-1)}...P^{(k+1)}\alpha^{(k)},$$

leading to the Gauss transform:

$$\tilde{M}^{(k)} = (I - \tilde{\alpha}^{(k)}e_k^T).$$

Thus (3.13) is equivalent to

$$\tilde{M}^{(n-1)}...\tilde{M}^{(1)}PA = U.$$

where $P = P^{(n-1)}...P^{(1)}$ is defined recursively by

$$P = I(\text{identity matrix}); \ P = P^{(}k) * P, \ k = 1 : n - 1.$$

Algorithm 3.2 Unscaled Partial Pivoting *LU* Decomposition

```
function [L,U,P]=myluunscaled(A,tol)
% Input: A, square matrix;
%           tol, tolerance to detect singularity of A
% Output: L, U unit lower and upper triangular matrices
%               P, permutation matrix
% Uunscaled partial pivoting strategy to get P*A=L*U
n=length(A);I=eye(n);P=I;
for k=1:n
    % Search for pivot row
    y=[zeros(k-1,1);A(k:n,k)];
    [p,ik]=max(abs(y));
     if p>tol
        A([k ik],:)=A([ik k],:);%Permute rows k and ik
        P([k ik],:)=P([ik k],:);%Update permutation matrix
        piv=1/A(k,k);
        A(k+1:n,k)=piv*A(k+1:n,k);
        A(k+1:n,k+1:n)=A(k+1:n,k+1:n)-A(k+1:n,k)*A(k,k+1:n);
    end
end
L=I+tril(A,-1);U=triu(A);
```

3.4.2   Scaled Partial Pivoting

In this case, we introduce "matrix row scales," whereby at the beginning of the computation, one stores the n scales:

$$s_i = \max\{|A(i, 1:n)|\}.$$

These scales are not changed throughout the computations. Thus, at the first reduction, the strategy of scaled partial pivoting seeks the row i_1, such that

$$|A(i_1, 1)|/s_{i_1} = \max_{i=1:n} |A(i, 1)/s_i|,$$

giving a relative maximum rather than an absolute one as is the case of unscaled partial pivoting. Obviously such implementation leads to major data transfers when performing row permutations. One way to avoid such tasks is through the introduction of an "index vector" that traces the permutations when they occur. The index vector is initially set to $1:n$. Permutations occur on the index vector and not on the matrix. We give such optimal storage version for the scaled partial pivoting strategy, indicating the number of flops for executing each of the steps.

Algorithm 3.3 Optimal Storage Algorithm for Scaled Partial Pivoting

```
function [L,U,P]=mylu(A,tol)
% Input: A a square matrix (invertible or not)
% Output: either:
%          L unit lower triangular
%          U upper triangular
%          P: Permutation matrix
%          P*A=L*U and in case
%          A singular display matrix rank
n=length(A);r=n;
P=eye(n);
% Compute the scales
s=max(abs(A'));s=s(:);
IV=1:n;
for k=1:n
    [p,ip]=max(abs(A(IV(k:n),k))./s(IV(k:n)));
    ip=ip+k-1;
    if abs(A(IV(ip),k))>tol
        IV([ip k])=IV([k ip]);
        piv=sign(A(IV(k),k))*p*s(IV(k));
        piv=1/piv;% 1 flop
        A(IV(k+1:n),k)=piv*A(IV(k+1:n),k);%(n-k) flops
        B=A(IV(k+1:n),k)*A(IV(k),k+1:n);%(n-k)^2 flops
        A(IV(k+1:n),k+1:n)=A(IV(k+1:n),k+1:n)-B;%(n-k)^2 flops
    else
        r=r-1;
    end
end
if r<n
    st=['matrix nearly singular with rank=' num2str(r)];
    disp(st);
end
U=triu(A(IV,:));L=eye(n)+tril(A(IV,:),-1);P=P(IV,:);
```

3.4.3 Solving a System $Ax = b$ Using the LU Decomposition

Given the $\{L, U, P\}$ decomposition of A, $PA = LU$, let $b \in \mathbb{R}^n$. In such case, the system $Ax = b$ is equivalent to $LUx = Pb$. Thus x can be obtained by solving successively two triangular systems:

$$Ly = Pb \text{ followed by } Ux = y. \tag{3.16}$$

3.5 MATLAB Commands Related to the LU Decomposition

1. The `lu` command.
   Such command operates under the following rule:
   `[L,U,P] = lu(A)` gives the matrices L, U and P (according to Proposition 3.5) such that `P*A = L*U`.

2. The `chol` command.
   Such command works out as follows.
   `R = chol(A)` gives an upper triangular matrix R such that $R^T R = A$.
   If A is not positive definite, an error message is printed to that effect.

3. The `backslash` command: $\backslash$ whereas `A\b` consists in an P, L, U decomposition followed in case A is invertible by solving a sequence of forward and backward substitutions as in (3.16).

3.6 Condition Number of a Square Matrix

Let $A \in \mathbb{R}^{n \times n}$ be an invertible matrix. For $b \in \mathbb{R}^n$, consider the system of linear equations,

$$Ax = b. \tag{3.17}$$

Let $x_c \in \mathbb{F}^n$ be a numerical solution obtained in applying an algorithm for solving (3.17). The condition number of A provides a relationship between $||x - x_c||/||x||$ and $||b - Ax_c||/||b||$.
This is shown by considering the following **regressive** analysis.
Starting with the identity:

$$x - x_c = A^{-1}(A(x - x_c)) = A^{-1}(b - Ax_c).$$

The above matrix norms property (including that of Frobenius) leads to:

$$||x - x_c||_p \leq ||A^{-1}||_p (||b - Ax_c||_p). \tag{3.18}$$

Note that when $p = 2$, the matrix norm may be $||A||_2$ or $||A||_F$. Furthermore, since $Ax = b$ gives $||b|| \leq ||A|| ||x||$, one has:

$$\frac{1}{||x||} \leq \frac{||A||}{||b||}.$$

Multiplying this inequality with (3.18) leads to:

$$\frac{||x - x_c||_p}{||x||} \leq cond_p(A) \frac{||b - Ax_c||_p}{||b||_p}, \ 1 \leq p \leq \infty. \tag{3.19}$$

where for a square invertible matrix $A \in \mathbb{R}^{n \times n}$, the condition number is defined as follows:

$$cond_p(A) = ||A||_p ||A^{-1}||_p.$$

From the above remark, note when $p = 2$ in the matrix norm we may use either $||A||_F$ or $||A||_2$ for matrix norms.

The `MATLAB` implementations of the condition number is done through the following syntax:

```
cond(A) returns the 2-norm condition number (the ratio of the
largest singular value of matrix A to its smallest).
Large condition numbers indicate a nearly singular matrix.
cond(A,p) returns the condition number of A in p-norm:
norm(A,p) * norm(inv(A),p),   p=1, 2, inf, or 'fro.'
```

3.7 Exercises

Exercise 3.1

Prove 1 in Proposition 3.1.

Exercise 3.2

Let $A, M \in \mathbb{R}^{n \times n}$ and consider the product $C = M \times A$ and the formula:

$$C(i, :) = \sum_{k=1}^{n} M(i, k) \times A(k, :), \; i = 1 : n. \tag{3.20}$$

1. Find and display using (3.20), the matrix $M \in \mathbb{R}^{n \times n}$ such that $M \times A$ results in the third row of A being changed so that it is added to a constant c times its first row (of A), with all other elements of A remaining unchanged, i.e.,

$$A(i, :) \leftarrow A(i, :), \; \forall i \neq 3 \text{ and } A(3, :) \leftarrow A(3, :) + c \times A(1, :).$$

2. Consequently, find using (3.20), the matrix $M \in \mathbb{R}^{n \times n}$ such that $M \times A$ results in the ith row of A being changed so that it is added to a constant c_i times its first row (of A), i.e.,

$$\forall i > 1, \; A(i, :) \leftarrow A(i, :) + c_i \times A(1, :).$$

Show that $M = I + ce_1^T$, where

$$c = \begin{pmatrix} 0 \\ c_2 \\ c_3 \\ \vdots \\ \vdots \\ c_n \end{pmatrix} \text{ and } e_1 = \begin{pmatrix} 1 \\ 0 \\ 0 \\ \vdots \\ \vdots \\ 0 \end{pmatrix}$$

Exercise 3.3 Algebra of Triangular Matrices

If L and M are square n by n lower triangular matrices, prove that

1. $L + M$ is lower triangular.

2. $N = L \times M$ is lower triangular.

3. In 2, show if $L_{ii} = M_{ii} = 1, \forall i$, then $N_{ii} = (L \times M)_{ii} = 1, \forall i$.

4. If L is a lower triangular invertible matrix, then L^{-1} is also lower triangular. Furthermore, if $L_{ii} = 1, \forall i$, then $L_{ii}^{-1} = 1, \forall i$.

Exercise 3.4

Prove that if a matrix A is pmp, then the matrix $A^{(1)} = M^{(1)}A$, where $M^{(1)}$ is the (lower triangular matrix) Gauss transform associated with the 1st column of A is such that $a_{22}^{(1)} \neq 0$.

Exercise 3.5

Prove that in (3.10) the elements of the diagonal in the matrix product $L_2^{-1}L_1$ are all equal to 1.

Exercise 3.6

Prove the uniqueness of Cholesky's decomposition.

Exercise 3.7

Prove formula (3.11).

Exercise 3.8 *Show, when the matrix A is symmetric and has the principal minor property (pmp), that $L = L_1$ in the LDL_1^T decomposition.*

Exercise 3.9 *Show that when the matrix A is symmetric and pmp the Gauss transforms $\{M^{(j)}|j = 1 : n - 1\}$ are such that the matrices $A^{(k)} = M^{(k)}...M^{(1)}A$ verify the property that $A^{(k)}(k : n, k : n)$ are also symmetric.*

Exercise 3.10 *Show that if A is **spd, symmetric and positive definite**, then all elements of the diagonal matrix D are positive.*

Exercise 3.11 *(Cholesky) Show that if A is **spd, symmetric and positive definite**, then there exists a unique lower triangular matrix B such that $A = BB^T$.*

3.8 Computer Exercises

Computer Exercise 3.1 *Transform the* `mylu1` *to become* `mylusym` *in a way that the operation:*

```
A(k+1:n,k+1:n)=A(k+1:n,k+1:n)-A(k+1:n,k)*A(k,k+1:n);
```

is done only on the lower part of the matrix A. Check out the operation count in the form of $O(mn^3)$. What is m?

Computer Exercise 3.2 *Modify the above* function `[L,D]=mylusym(A)` *to obtain a "simple"* MATLAB *implementation of Cholesky decomposition:* function `B=mychol(A)`.

Computer Exercise 3.3 *Complete the direct Cholesky's decomposition* function `B=mycholdir(A)` *of Algorithm 3.4 which computes* B *directly from* A *using external products. Find the number of flops.*

Algorithm 3.4 A Direct Cholesky's Decomposition for an spd Matrix

```
function B=mycholdir(A)
n=length(A);
for k=1:n
    A(k,k)=sqrt(A(k,k));% 1 sqrt O(1)
    c=1/A(k,k); %1 flop
    A(k+1:n,k)=c*A(k+1:n,k); % n-k flops
    for j=k+1:n
        .....................................;%<== Complete here
    end
end
B=tril(A);
```

Based on the two previous exercises and using the MATLAB functions tic toc, construct **random diagonally dominant symmetric positive definite matrices** of sizes $n = 64, 128, 256, 512, 1024$ and construct a table that gives the elapsed execution time for:

1. MATLAB chol

2. mychol

3. mycholdir

Computer Exercise 3.4 Finding the inverse of a matrix

Let A be an n by n pmp matrix on which one applies Naive Gauss Reduction. Let also I be the n by n identity matrix. Complete Algorithm 3.5 that solves $AB = I$, through a minimum number of elementary matrix operations (and flops) to get $B = A^{(-1)}$.

Algorithm 3.5 Algorithm for the Inverse of a Square pmp Matrix

```
function B=myinv(A)
% Input:    A pmp square matrix
% Output:   B inverse of A
% Method:   Applies Gauss Reduction followed
%           by back substitution
%           to solve A*B=I
n=length(A);
B=eye(n);
for k=1:n-1
   piv=A(k,k);
   A(k+1:n,k)=(1/piv)*A(k+1:n,k);
%    A(k+1:n,k+1:n)=.............................;
%    B(k+1:n,1:k-1)=.............................;
%    B(k+1:n,k)=....................;
end
for k=n:-1:1
    %B(k,:)=....................;
    %B(1:k-1,:)=..............................;
end
```

---

1. Find $f(n)$, the number of flops necessary to compute B using Algorithm 3.5.

2. Test the resulting `myinv` vs `inv` function on $n \times n$ randomly generated matrices $A = A(n)$, with sizes $n = 8, 16, 32, 64, 128, 256$, giving your results in the form of a table that indicates in its columns the variations with respect to n of the following variables:
   - $T1(n)$: The elapsed time to execute `myinv`
   - $T2(n)$: The elapsed time to execute `inv`
   - $\|B - C\|$: where B=myinv(A); C=inv(A)
   - $\frac{T1(n)}{T2(n)}$
   - $f(n)$
   - $\frac{T1(n)}{f(n)}$
   - $\frac{T2(n)}{f(n)}$

Chapter 4

Orthogonal Factorizations and Linear Least Squares Problems

The purpose of linear least squares approaches is mainly to approximate solutions of overdetermined systems, i.e., sets of equations in which there are more equations than unknowns.

4.1 Formulation of Least Squares Problems: Regression Analysis

Specifically, consider a matrix $A \in \mathbb{R}^{m \times n}$, and let $y \in \mathbb{R}^m$. The least squares solution $w \in \mathbb{R}^n$ of the linear system $Aw\text{ "}=\text{ "}y$ is defined as:

$$\text{Find } w \in \mathbb{R}^n, \text{ such that } \|y - Aw\| = \min_{z \in \mathbb{R}^n} \|y - Az\|, \quad (4.1)$$

where for $x \in \mathbb{R}^n$, $\|x\| = \|x\|_2 = \sqrt{x^T x}$ is the l_2 vector norm. The existence and uniqueness of such a solution will be discussed in Sections 4.2 and 4.3. We first give an application of linear least squares problems in statistical computations.

4.1.1 Least Squares and Regression Analysis

The most important application of Linear Least Squares is in data fitting. The best fit in the least squares sense minimizes the sum of squared residuals, a residual being the difference between an observed value and the fitted value provided by a model. A well-known example of linear least squares problems is obtained when dealing with a set of statistical data:

$$D_n = \{(x_i, y_i) \in \mathbb{R}^2, \ i = 1, .., n\}.$$

Linear regression data fit seeks an ordered pair $w = (\alpha, \beta) \in \mathbb{R}^2$, such that:

$$\sum_{i=1}^{n} (\alpha x_i + \beta - y_i)^2 = \min_{z=(a,b) \in \mathbb{R}^2} \sum_{i=1}^{n} (ax_i + b - y_i)^2. \quad (4.2)$$

Similarly in **quadratic regression**, one looks for the triplet $w = (\alpha, \beta, \gamma) \in \mathbb{R}^3$, such that:

$$\sum_{i=1}^{n} (\alpha x_i^2 + \beta x_i + \gamma_i - y_i)^2 = \min_{z=(a,b,c)\in\mathbb{R}^3} \sum_{i=1}^{n} (ax_i^2 + bx_i + c - y_i)^2. \quad (4.3)$$

4.1.2   Matrix Formulation of Regression Problems

In the linear case, let:

$$A = \begin{pmatrix} x_1 & 1 \\ x_2 & 1 \\ .. & .. \\ x_{n-1} & 1 \\ x_n & 1 \end{pmatrix} \text{ and } y = \begin{pmatrix} y_1 \\ y_2 \\ . \\ y_{n-1} \\ y_n \end{pmatrix}, \quad (4.4)$$

then, (4.2) is equivalent to:

$$\|Aw - y\|^2 = \min_{z\in\mathbb{R}^2} \|Az - y\|^2.$$

Similarly, in the quadratic case, one verifies that for:

$$A = \begin{pmatrix} x_1^2 & x_1 & 1 \\ x_2^2 & x_2 & 1 \\ .. & .. \\ x_{n-1}^2 & x_{n-1} & 1 \\ x_n^2 & x_n & 1 \end{pmatrix} \quad (4.5)$$

the solution to (4.3) is such that:

$$\|Aw - y\|^2 = \min_{z\in\mathbb{R}^3} \|Az - y\|^2.$$

4.2   Existence of Solutions Using Quadratic Forms

Obviously, (4.1) is equivalent to:

$$\|y - Aw\|^2 = \min_{z\in\mathbb{R}^n} \|y - Az\|^2.$$

Note that:

$$\|y - Az\|^2 = (y - Az)^T(y - Az) = \|y\|^2 + z^T B z - 2z^T A^T y, \quad (4.6)$$

where $B = A^T A \in \mathbb{R}^{n \times n}$ is symmetric and positive semi-definite matrix, given that $(A^T A)^T = A^T A$ and $z^T B z = \|Az\|^2 \geq 0$. Since $\|y\|$ is independent from z then, using (4.6), one notes that minimizing $\|y - Az\|^2$ reduces to that of minimizing over $\mathbb{R}^n$,

$$\phi(z) = \frac{1}{2} z^T B z - z^T c, \text{ where } c = A^T y \in \mathbb{R}^n. \tag{4.7}$$

i.e., finding:

$$\phi(w) = \min_{z \in \mathbb{R}^n} \phi(z). \tag{4.8}$$

We prove now a classical result for quadratic forms of the type (4.7).

Theorem 4.1 $w \in \mathbb{R}^n$ *solves the least squares problem (4.1) if and only if it solves the minimization problem (4.8) which is also equivalent to w solving the normalized system:*

$$Bw = c \iff A^T A w = A^T y. \tag{4.9}$$

Proof. As:

$$\phi(z) = \frac{1}{2} \Big(\sum_{i=1}^{n} B_{ii} z_i^2 + \sum_{i=1}^{n} z_i \sum_{j \neq i} z_j (B_{ij} + B_{ji}) \Big) - \sum_{i=1}^{n} z_i c_i,$$

one notes that $\phi : \mathbb{R}^n \to \mathbb{R}$ is a quadratic functional. Hence, on the basis that the matrix B is symmetric ($B_{ij} = B_{ji}$), a necessary condition for $\phi(z) \mid z \in \mathbb{R}^n$ to have a minimum $\phi(w)$ is that:

$$\frac{\partial \phi}{\partial z_i}(w) = 0, \forall i = 1, ..., n, \iff \sum_{i=j}^{n} B_{ij} w_j = c_i, \forall i = 1, ..., n. \tag{4.10}$$

Thus, any solution $w \in \mathbb{R}^n$ of the least squares problem (4.1) must solve (4.9). Conversely, if $w \in \mathbb{R}^n$ solves (4.9), i.e., verifies $Bw = c$, then,

$$\forall z \in \mathbb{R}^n : \phi(z) - \phi(w) = \frac{1}{2}(z^T B z - w^T B w) - (z - w)^T B w.$$

Hence,

$$\forall z \in \mathbb{R}^n : \phi(z) - \phi(w) = \frac{1}{2}(z - w)^T B(z + w) - (z - w)^T B w$$

$$= \frac{1}{2}(z - w)^T B(z - w) \geq 0. \qquad \blacksquare$$

As a consequence of this result, given that $B \in \mathbb{R}^{n \times n}$ so that $\mathrm{rank}(B) \leq n$ and since

$$r = \mathrm{rank}(B) = \mathrm{rank}(A^T A) = \mathrm{rank}(A) \leq \min(m, n),$$

then one of the following two situations may occur:

1. $m < n$ (A has less rows than columns): then $r = \text{rank}(B) = \text{rank}(A) \le \min(m, n) < n$ and the system (4.9) has several solutions given that B is rank deficient and A is also column rank deficient.

2. $m \ge n$ (The number of columns in A is at most its number of rows): then $r = \text{rank}(A) \le \min(m, n) \le n$. In such situation, two cases can occur:

   (a) $r = n$, (A full column rank), wherein the least squares problem has a unique solution. Note in this case that B is positive definite matrix as $z^T B z = ||Az||^2 = 0$ implies $z = 0$ if and only if $r = \text{rank}(A) = n$.

   (b) $r < n$ in which, likewise the first case ($m < n$), the least squares problem may have several solutions given that B is rank deficient and A column rank deficient.

Hence, based on Theorem 4.1, one concludes from case 2(a) (full column rank matrix A), the following result:

Corollary 4.1 *If $A \in \mathbb{R}^{m \times n}$ with $n \le m$ and the rank of A is n (full rank case), then $B = A^T A$ is symmetric and positive definite and the least squares problem (4.1) has a unique solution $z \in \mathbb{R}^n$, that solves the square normalized linear symmetric system:*

$$Bz = c, \text{ i.e., } A^T A z = A^T y, \tag{4.11}$$

■

4.2.1 Full Rank Cases: Application to Regression Analysis

To illustrate the application to full rank cases, consider the linear regression example and note that:

$$B = \begin{pmatrix} \sum_{i=1}^{n} x_i^2 & \sum_{i=1}^{n} x_i \\ \sum_{i=1}^{n} x_i & n \end{pmatrix} = \begin{pmatrix} b_{11} & b_{12} \\ b_{12} & b_{22} \end{pmatrix} \tag{4.12}$$

with

$$c = \begin{pmatrix} \sum_{i=1}^{n} x_i y_i \\ \sum_{i=1}^{n} y_i \end{pmatrix} = \begin{pmatrix} c_1 \\ c_2 \end{pmatrix}$$

and for the quadratic regression case (4.3):

$$B = \begin{pmatrix} \sum_{i=1}^{n} x_i^4 & \sum_{i=1}^{n} x_i^3 & \sum_{i=1}^{n} x_i^2 \\ \sum_{i=1}^{n} x_i^3 & \sum_{i=1}^{n} x_i^2 & \sum_{i=1}^{n} x_i \\ \sum_{i=1}^{n} x_i^2 & \sum_{i=1}^{n} x_i & n \end{pmatrix} = \begin{pmatrix} b_{11} & b_{12} & b_{13} \\ b_{12} & b_{22} & b_{23} \\ b_{13} & b_{23} & b_{33} \end{pmatrix} \tag{4.13}$$

with

$$c = \begin{pmatrix} \sum_{i=1}^{n} x_i^2 y_i \\ \sum_{i=1}^{n} x_i y_i \\ \sum_{i=1}^{n} y_i \end{pmatrix} = \begin{pmatrix} c_1 \\ c_2 \\ c_3 \end{pmatrix}$$

One easily verifies the following:

1. If the data $\{x_i\}$ in D_n are such that $x_i \neq x_j$ for at least one $i \neq j$, then, in (4.4) rank(A) = 2 (as one can extract from A a 2 by 2 Vandermonde submatrix), implying that in (4.12), B is symmetric and positive definite. Similarly,

2. If the data $\{x_i\}$ in D_n are such that there are at least three indices i, j, and k such that the corresponding data are all distinct, then, in (4.5), rank(A) = 3 (as one can extract from A a 3 by 3 Vandermonde submatrix), implying that in (4.13), B is also symmetric and positive definite.

Hence both (4.2) and (4.3) have unique solutions that can be found using **Cholesky decomposition** $B = LL^T$ followed by solving successively: $Lz_1 = c$ (forward substitution) and $L^T z = z_1$ (backward substitution). In the sequel, we shall show that it is numerically preferable to use the QR factorization of A rather than the Cholesky factorization of $A^T A$.

4.3 Existence of Solutions through Matrix Pseudo-Inverse

In the previous section, a first discussion of the existence of solutions to the linear least squares problem was sought using quadratic forms. We consider now an alternate approach that allows obtaining a solution x in $\mathbb{R}^n$ for the general least squares problem (4.1), given by a matrix $A \in \mathbb{R}^{m \times n}$ and a vector $y \in \mathbb{R}^m$.

As was shown in Section 4.2, any solution x to (4.1), solves the normalized problem (4.11):

$$A^T A x = A^T y. \tag{4.14}$$

Recall that $(A^T A)$ is positive definite (spd) if and only if A is of full column rank, i.e., rank(A) = n; in this situation, the solution of (4.11) is unique. We now study the general situation. Let the set of solutions be denoted by

$$S(A, y) = \{x \in \mathbb{R}^n | x \text{ solves the least squares problem (4.1)}\}.$$

To characterize this set, we start by proving the following result.

Theorem 4.2 *For any $x \in \mathbb{R}^n$, the 2-norm of the residual $r = y - Ax$ is minimum if and only if $A^T r = 0$, i.e., $r \in Null(A^T) = (Im(A))^\perp$.*

Proof. Let us denote the range of A by:

$$Im(A) = \{z \in \mathbb{R}^m \mid \exists x \in \mathbb{R}^n, \ y = Az\}.$$

Since $\mathbb{R}^m = \text{Im}(A) \oplus \text{Im}(A)^\perp$, where $\text{Im}(A)^\perp$ is the orthogonal complement of $\text{Im}(A)$ in $\mathbb{R}^m$, we can express $y = c + d$ where $c \in \text{Im}(A)$ and $d \in \text{Im}(A)^\perp$. Therefore, for any $x \in \mathbb{R}^n$,

$$
\begin{aligned}
\|r\|^2 &= \|y - Ax\|^2, \\
&= \|c - Ax\|^2 + \|d\|^2,
\end{aligned}
$$

since $c \in \text{Im}(A)$. This sum of norms is minimum if and only if x is such that $Ax = c$ and therefore $r = d$, i.e., $r \in (\text{Im}(A))^\perp = \text{Null}(A^T)$, implying $A^T r = 0$. ∎

As a consequence, we obtain now existence of solutions to the linear least squares problem (4.1).

Theorem 4.3 *The set $\mathcal{S}(A, y)$ is non-empty.*

Proof. The proof is constructive in obtaining $\mathcal{S}(A, y)$. By considering c, the orthogonal projection of $y = c + r \in \mathbb{R}^m$ onto $\text{Im}(A)$, then any element $x \in A^{-1}\{c\} = \{z \in \mathbb{R}^n, \ c = Az\}$, solves (4.1) or equivalently (4.14) as the set $A^{-1}\{c\}$ is non-empty. Furthermore for any $x \in A^{-1}\{c\}$, one has $A^T Ax = A^T c = A^T(y - r) = A^T y$, i.e., x solves (4.9). ∎

In the situation of a rank deficient problem, i.e., when the matrix $(A^T A)$ is singular, this result shows that $A^T y \in \text{Im}(A^T A)$ else the system (4.14) would be inconsistent. Furthermore, such element is unique whenever A is of full rank, i.e., when $(A^T A)$ is spd. This is summarized as follows:

Theorem 4.4 *The non-empty set $\mathcal{S}(A, y)$ is restricted to only one vector if and only if the columns of A are independent.*

We introduce now the notion of the pseudo-inverse of a matrix.

Theorem 4.5 (Definition of the Moore-Penrose Pseudo-Inverse)
There exists a unique vector $x \in \mathcal{S}(A, y)$, which norm is minimum. This vector is orthogonal to the null space of A and is denoted by:

$$
x = A^+ y. \tag{4.15}
$$

The matrix A^+ is called generalized inverse or Moore-Penrose pseudo-inverse.

Proof. When the mapping $x \to Ax$ is a one-to-one mapping (i.e., $\dim \text{Im}(A) = n$), the property is obvious. Otherwise, given $y = c + r$, $c \in \text{Im}(A)$, $r \in (\text{Im}(A))^\perp$, let $x_0, x_1 \in \mathbb{R}^n$ be 2 distinct elements in $A^{-1}\{c\}$. Then obviously $x_1 - x_0 \in \text{Null}(A)$. Furthermore, given that $\mathbb{R}^n = \text{Null}(A) \oplus \text{Null}(A)^\perp$, then one writes $x_i = e_i + f_i$, $e_i \in \text{Null}(A)$ and $f_i \in \text{Null}(A)^\perp$, $i = 0, 1$. As $x_0 - x_1 \in \text{Null}(A)$ implies that $f_0 = f_1 = f$, then one concludes that for any $x \in A^{-1}\{c\}$, there exists a unique $f \in (\text{Null}(A))^\perp$, such that:

$$
x = \Pi_{\text{Null}(A)}(x) + f = c \text{ with } \|x\|^2 = \|\Pi_{\text{Null}(A)}(x)\|^2 + \|f\|^2,
$$

where $\Pi_{\mathrm{Null}(A)}(x)$ is the orthogonal projection of x onto $\mathrm{Null}(A)$. Obviously $||x||$ is minimal if and only if one takes $x = f$, with such choice being unique. ∎

Thus, given $A \in \mathbb{R}^{m \times n}$, $y \in \mathbb{R}^m$, the steps to be followed to get the unique $x = A^+ y$ are as follows:

$$
\begin{aligned}
&\text{(i) Get } c = \Pi_{\mathrm{Im}(A)} y. \\
&\text{(ii) Get any } w \in A^{-1}\{c\}. \\
&\text{(iii) Take } x = \Pi_{(\mathrm{Null}(A))^\perp} w.
\end{aligned}
\tag{4.16}
$$

Then $x = A^+ y$. Step (iii) is skipped when A is of full-column rank.

4.3.1 Obtaining Matrix Pseudo-Inverse through Singular Value Decomposition

Let $r \leq \min(m, n)$ be the rank of $A \in \mathbb{R}^{m \times n}$; the case $r < n$ corresponds to the rank deficient linear least squares problems in which case there is an infinite set of solutions for (4.1). Recall that for any solution x, the set of all solutions is expressed by $\mathcal{S} = x + (A)$ where (A) is the $(n - r)$-dimensional null space of A. The Singular Value Decomposition (SVD) (see 2.3.2) of A is given by:

$$
A = USV^T,
$$

that can be rewritten as:

$$
A = U \begin{pmatrix} S_r & 0_{r,n-r} \\ 0_{m-r,r} & 0_{m-r,n-r} \end{pmatrix} V^T,
\tag{4.17}
$$

where the diagonal entries of the matrix $S_r = \mathrm{diag}(\sigma_1, \cdots, \sigma_r) \in \mathbb{R}^{r \times r}$ are positive, $U \in \mathbb{R}^{m \times m}$, $V \in \mathbb{R}^{n \times n}$ are orthogonal matrices.

Theorem 4.6 *Let the matrix $A \in \mathbb{R}^{m \times n}$ be of rank $r \leq \min(m, n)$ and $r < n$ with its SVD defined by (4.17). Then its generalized inverse of A as characterized by Theorem 4.5, is given by:*

$$
A^+ = V \begin{pmatrix} S_r^{-1} & 0_{m-r,r} \\ 0_{r,n-r} & 0_{n-r,r} \end{pmatrix} U^T.
\tag{4.18}
$$

Proof. The columns of U and V are separated into two sets of columns: $U = [U_1, U_2]$ and $V = [V_1, V_2]$, where U_1 and V_1 are defined by r columns. The SVD exhibited in (4.17) can then be transformed into the compact (economy) form:

$$
A = U_1 S_r V_1^T.
\tag{4.19}
$$

Let us now follow the procedure described in (4.16) for computing $A^+ y$ where y is any vector of $\mathbb{R}^m$. Since the columns of U_1 form an orthonormal basis of

$\mathrm{Im}(A)$, the orthogonal projection of y onto this subspace can be expressed by $c = \Pi_{\mathrm{Im}(A)}y = U_1 U_1^T y$. Let $w \in \mathbb{R}^n$ be any solution such that $Aw = c$; it is not necessary to build it explicitly but only $x = \Pi_{(\mathrm{Null}(A))^\perp} w$. Therefore from (4.19), we may write the following derivation:

$$
\begin{aligned}
U_1 S_r V_1^T w &= U_1 U_1^T y, \\
S_r V_1^T w &= U_1^T y, \\
V_1^T w &= S_r^{-1} U_1^T y, \\
V_1 V_1^T w &= V_1 S_r^{-1} U_1^T y, \\
x = \Pi_{(\mathrm{Null}(A))^\perp} w &= V_1 S_r^{-1} U_1^T y.
\end{aligned}
$$

The last equality is true since V_1 is orthogonal to V_2 which is a basis of $\mathrm{Null}(A)$. It proves the property (4.18). ∎

Practically, the solution $x = A^+ y$ is obtained by $c = U^T y = \begin{pmatrix} c_1 \\ c_2 \end{pmatrix}$ where $c_1 \in \mathbb{R}^r$, and then $x = V_1 S_r^{-1} c_1$.

One checks the following properties of A^+:

$$
A^+ \in \mathbb{R}^{n \times m}, \ AA^+ A = A \text{ and } A^+ AA^+ = A^+.
$$

In MATLAB the pseudo-inverse is implemented through the command `pinv`. For a rectangular matrix $A \in \mathbb{R}^{m \times n}$, the vector $x = A\backslash y$ is such that the norm of $r = y - Ax$ is minimum. When the matrix is of full column rank, the solution is unique and $\mathrm{pinv}(A) * y = A\backslash y$. When the matrix is not of full column rank, $\mathrm{pinv}(A) * y$ is the solution of minimum norm whereas $A\backslash y$ is usually not (see Section 4.8.1).

The following are special cases of pseudo-inverses.

Theorem 4.7 (Overdetermined Least Squares Problem) *When $m \geq n$ and $A \in \mathbb{R}^{m \times n}$ is of full (column) rank $r = n$, the pseudo-inverse is given by:*

$$
A^+ = (A^T A)^{-1} A^T. \tag{4.20}
$$

Proof. The result is a direct consequence of the normal system (4.14) which matrix is non-singular. ∎

Corollary 4.1 *When A is a square non-singular matrix, then $A^+ = A^{-1}$.*

This corollary proves that the generalized inverse extends the definition of an inverse to rectangular matrices.

Expression (4.20) provides a method for solving a full column rank overdetermined least squares linear system through the normal equation (4.14). The computation involves a matrix multiplication and then a Cholesky factorization. That normal system (4.14) is one which condition number (see the definition of the condition number in Section 3.6):

$$cond_2(A^T A) = \|A^T A\| \, \|(A^T A)^{-1}\| = \frac{\sigma_1^2}{\sigma_n^2}, \tag{4.21}$$

where $(\sigma_i)_{1,n}$ are the singular values of A ordered in descending order. If we consider the expression obtained in (4.18) from the singular value decomposition (SVD) of A, the condition number of the system solved with S_n is only $\frac{\sigma_1}{\sigma_n}$, which can be a much lower value. In Section 4.6, another method is given which reaches this goal and which involves less operations.

Finally, in the case of rank deficient least squares problems, pseudo-inverses involve an expensive computation using the SVD of a matrix A, although the economy version is used as in (4.19):

$$x = A^+ y = V_1 S_r^{-1} U_1^T y. \tag{4.22}$$

To bypass either normal systems or SVDs when solving least squares problems, we turn now to QR decompositions of a matrix.

4.4 The QR Factorization Theorem

One alternative to solving the least squares problem without computing the normal system (4.11) is to use a decomposition of the matrix A which is another important factorization in Computational Linear Algebra (as the LU factorization in Chapter 3). It applies to any rectangular matrix:

Theorem 4.8 (QR Factorization) *Let $A \in \mathbb{R}^{m \times n}$, then there exists an orthogonal matrix $Q \in \mathbb{R}^{m \times m}$ and an upper triangular matrix $R \in \mathbb{R}^{m \times n}$, such that:*

$$A = QR, \tag{4.23}$$

To prove the existence of this factorization, one may proceed in defining the orthogonal matrix Q of (4.23), as a product of elementary orthogonal transforms using (i) **Householder** reflectors, (ii) **Givens** rotations, or in obtaining the orthogonal basis by orthogonal projections by the (iii) **Gram-Schmidt** process.

4.4.1 Householder Transforms

The main idea of a Householder transform stems from matrices obtained by modifying the identity matrix by a rank one external product matrix as introduced in (3.2) in Chapter 3. Specifically, if in

$$M = M(u, v) = I - uv^T,$$

we consider $v \neq 0$ and take $u = \frac{2}{v^T v} v$, $M(u, v)$ becomes the Householder matrix, namely:

$$H = H(v) = I - \frac{2}{v^T v} v v^T. \qquad (4.24)$$

Then, for a non-zero vector $w \in \mathbb{R}^m$, $w \neq 0$ that is not collinear to $e_1 = (1, 0, ..., 0)^T$ (i.e., there exists $i \geq 2$ such that $w_i \neq 0$), we seek the vector $v \in \mathbb{R}^n$, such that the associated Householder transform $H = H(v)$ transforms w into a vector Hw that is co-linear to the unit vector $e_1 = (1\, 0\, 0 ... 0)^T \in \mathbb{R}^n$, i.e.,

$$Hw = w - \frac{2v^T w}{v^T v} v = \gamma e_1, \ \gamma \in \mathbb{R}. \qquad (4.25)$$

Note that v cannot be orthogonal to w, otherwise $Hw = w$. Thus, on the basis that $v^T w \neq 0$, the last identity proves that v must belong to the subspace generated by the two linearly independent vectors, $\{w, e_1\}$, i.e., there exists $c \in \mathbb{R}$, such that

$$v = w + c e_1.$$

To obtain c, we substitute v in (4.25). One has first:

$$v^T w = ||w||^2 + c w_1 \text{ and } v^T v = ||w||^2 + c^2 + 2 c w_1,$$

and therefore,

$$Hw = w - \frac{2(||w||^2 + c w_1)}{||w||^2 + c^2 + 2 c w_1} (w + c e_1).$$

Thus:

$$Hw = (1 - \frac{2(||w||^2 + c w_1)}{||w||^2 + c^2 + 2 c w_1}) w - \frac{2c(||w||^2 + c w_1)}{||w||^2 + c^2 + 2 c w_1} e_1.$$

Hence:

$$Hw = \frac{c^2 - ||w||^2}{||w||^2 + c^2 + 2 c w_1} w - \frac{2c(||w||^2 + c w_1)}{||w||^2 + c^2 + 2 c w_1} e_1. \qquad (4.26)$$

Therefore, Hw is collinear to e_1 if and only if

$$c = \pm ||w||,$$

which means that the problem of determining v from w has two solutions that belong to the span of $\{w, e_1\}$, given by:

$$v = w \pm ||w|| e_1, \qquad (4.27)$$

thus leading to $v^T w = ||w||^2 + \pm ||w|| w_1 = ||w|| (||w|| \pm w_1)$. As w is not collinear to e_1, $||w|| \pm w_1 \neq 0$. However, to avoid loss of significant figures, we take:

$$c = \text{sign}(w_1) ||w||,$$

where for $x \in \mathbb{R}$, $\mathrm{sign}(x) = 1$ for $x \geq 0$ and $\mathrm{sign}(x) = -1$ for $x < 0$. Thus:

$$v = w + \mathrm{sign}(w_1)\|w\|e_1, \tag{4.28}$$

and from (4.26) one obtains:

$$Hw = -\,\mathrm{sign}(w_1)\|w\|e_1, \tag{4.29}$$

Thus, H annihilates all the components of w, except the first one. If:

$$w = \begin{pmatrix} w_1 \\ w_2 \\ \dots \\ w_n \end{pmatrix},$$

then:

$$Hw = w - 2\frac{v^T w}{v^T v}v = \begin{pmatrix} -\,\mathrm{sign}(w_1)\|w\| \\ 0 \\ \dots \\ 0 \end{pmatrix}.$$

On the other hand, the matrix H satisfies the following properties:

Proposition 4.1 *The matrix $H = I - \frac{2}{v^T v}vv^T$ is symmetric and orthogonal. Therefore $H^T = H^{-1} = H$ and $H^2 = I$. Every vector of the hyperplane $(v)^{\perp}$ is invariant by H and $Hv = -v$: H is the orthogonal reflector w.r.t. the hyperplane $(v)^{\perp}$.*

Proof. We leave it to the reader to give the details of the proof.   ∎

Let us introduce now the following notations:

$$\alpha = \|w\| \text{ and } \beta = \frac{2}{v^T v}. \tag{4.30}$$

Using the identities:

$$v^T w = \|w\|^2 + \mathrm{sign}(w_1)w_1\|w\| = \|w\|^2 + |w_1|\|w\|,$$

and

$$v^T v = (w + \mathrm{sign}(w_1)\|w\|e_1)^T(w + \mathrm{sign}(w_1)\|w\|e_1) = 2\|w\|^2 + 2|w_1|\|w\|,$$

i.e.,

$$v^T v = 2\|w\|(\|w\| + |w_1|),$$

one obtains the relation:

$$\beta = \frac{2}{v^T v} = \frac{1}{\|w\|(\|w\| + |w_1|)}, \text{ i.e., } \beta = \frac{1}{\alpha(\alpha + |w_1|)}. \tag{4.31}$$

On the other hand, one checks that, for any vector $u \in \mathbb{R}^m$,

$$Hu = u - 2\frac{v^T u}{v^T v}v. \tag{4.32}$$

This implies that the multiplication Hu involves $O(m)$ operations (one scalar product and one **saxpy**), whereas the regular matrix-vector multiplication involves $O(m^2)$ operations.

The above results can be generalized to a vector $w \in \mathbb{R}^n$, such that the extracted subvector, $w^{(k)} \in \mathbb{R}^n$, given by:

$$w^{(k)} = \begin{pmatrix} 0 \\ . \\ 0 \\ w_k \\ w_{k+1} \\ . \\ . \\ . \\ w_n \end{pmatrix}$$

is non-zero (i.e., $w^{(k)} \neq 0$). In that case, one computes $v^{(k)} = w^{(k)} + \text{sign}(w_k)\|w^{(k)}\|e_k \neq 0$, such that $H^{(k)} = I - \frac{2}{(v^{(k)})^T v^{(k)}}v^{(k)}(v^{(k)})^T$ satisfies:

$$(H^{(k)}w)_i = 0, \quad \forall i > k,$$

specifically:

$$H^{(k)}w^{(k)} = \begin{pmatrix} 0 \\ . \\ 0 \\ -\text{sign}(w_k)\|w^{(k)}\| \neq 0 \\ 0 \\ . \\ 0 \end{pmatrix}$$

Following the argument used for $k = 1$, one obtains the generalized Householder vector:

$$v^{(k)} = \begin{pmatrix} 0 \\ . \\ 0 \\ w_k + \text{sign}(w_k)\|w^{(k)}\| \\ w_{k+1}. \\ w_n \end{pmatrix}$$

Letting:

$$\alpha^{(k)} = \|w^{(k)}\| \text{ and } \beta^{(k)} = \frac{2}{(v^{(k)})^T v^{(k)}}, \tag{4.33}$$

one shows then, that

$$H^{(k)}w = \begin{pmatrix} w_1 \\ \cdot \\ w_{k-1} \\ -\operatorname{sign}(w_k)\alpha^{(k)} \\ 0 \\ \cdot \\ 0 \end{pmatrix}$$

with a similar identity to (4.31) between $\alpha^{(k)}$ and $\beta^{(k)}$, namely:

$$\beta^{(k)} = \frac{2}{(v^{(k)})^T v^{(k)}} = \frac{1}{\|w^{(k)}\|(\|w^{(k)}\| + |w_k|)},$$

i.e.,

$$\beta^{(k)} = \frac{1}{\alpha^{(k)}(\alpha^{(k)} + |w_k|)}. \tag{4.34}$$

4.4.2 Steps of the QR Decomposition of a Matrix

Consider the matrix $A = [a^1, a^2, \cdots, a^n] \in \mathbb{R}^{m \times n}$ where a^i denotes the i-th column. Let $r = \min(m-1, n)$. We then construct a sequence of Householder matrices $\{H^{(1)}, H^{(2)}, ..., H^{(r)}\}$ such that $H^{(r)}...H^{(1)}A$ becomes an upper triangular matrix.

Let $H^{(1)}$ be the first Householder transform corresponding to $w = a^1$. Define then

$$A^{(1)} = H^{(1)}A = [H^{(1)}a^1, H^{(1)}a^2, \cdots, H^{(1)}a^n].$$

Thus:

$$H^{(1)}A = \begin{pmatrix} * & * & * & \cdot & * \\ 0 & * & * & \cdot & * \\ \cdot & \cdot & \cdot & \cdot & \cdot \\ 0 & * & * & \cdot & * \\ 0 & * & * & \cdot & * \end{pmatrix}$$

Hence, by recurrence, one obtains an upper triangular matrix A_r from:

$$R = A^{(r)} = H^{(r)} \cdots H^{(1)} A. \tag{4.35}$$

At this point, the definition of r becomes clear since, if $m > n$ then, there are n Householder transformations or else, only the $m-1$ first columns define Householder transformations. If we let $Q = H^{(1)}H^{(2)}, \cdots, H^{(r)}$, then obviously $QQ^T = I$ and $A = QR$. ∎

Remark 4.1 *The matrix $H = H(w)$ depends on the vector v, which computation (using formula (4.28)) requires finding $\sum_{i=1}^{n} w_i^2$. This could lead to* *overflows. To avoid such problem, one introduces* $\tilde{w} = \frac{1}{\|w\|_\infty} w$ *and uses the property:*

$$H = H(w) = H(\tilde{w}).$$

Since $|\tilde{w}_i| \leq 1, \forall i$, *this means that the computation of* $\|\tilde{w}\|$ *would be less sensitive to overflows than* $\|w\|$.

Thus, for $w^{(k)}$, to avoid overflow, one may use the vector:

$$\tilde{w}^{(k)} = \frac{1}{\|w^{(k)}\|_\infty} w^{(k)},$$

so that the computation of $\alpha^{(k)}$ in (4.33) uses $\|\tilde{w}^{(k)}\|$. In the case when $\alpha^{(k)} = 0$, it is obvious that $H^{(k)} = I$. Otherwise, given $\alpha^{(k)} > 0$, one computes $\beta^{(k)}$ using the identity:

$$(v^{(k)})^T v^{(k)} = 2\alpha^{(k)}(\alpha^{(k)} + |\tilde{w}_k^{(k)}|).$$

This results in:

$$\beta^{(k)} = \frac{1}{\alpha^{(k)}(\alpha^{(k)} + |\tilde{w}_k^{(k)}|)}.$$

As for the computation of $H^{(k)}Z$, where $Z \in \mathbb{R}^{m \times n}$, given the structure of $v^{(k)}$, it reduces to modifying only the submatrix $Z(k:m, k:n)$:

$$Z(k:m, k:n) := Z(k:m, k:n) - \beta^{(k)} v^{(k)}(k:m)((v^{(k)}(k:m))^T Z(k:m, k:n)).$$

4.4.3 Particularizing When $m > n$

This is the situation when solving an overdetermined linear least squares problem.

Proposition 4.2 (Economy Version of QR) *When* $m > n$, *the matrices* Q *and* R *can be expressed by* $Q = [Q_1, Q_2]$ *and* $R = \begin{pmatrix} \tilde{R} \\ 0 \end{pmatrix}$, *where* $Q_1 \in \mathbb{R}^{m \times n}$ *and* $\tilde{R} \in \mathbb{R}^{n \times n}$. *Therefore*

$$A = Q_1 \tilde{R}. \tag{4.36}$$

When A *is of full column rank, the matrix* $\tilde{R}$ *is a non-singular triangular square matrix.*

When necessary for subsequent calculations, the sequence of the vectors $(w_k)_{1,n}$, can be stored for instance in the lower part of the transformed matrix $A^{(k)}$. One can also compute explicitly the matrix Q_1 of (4.36).

Proposition 4.3 (Operation Counts (Householder QR)) *When* $m > n$, *the computation for obtaining* R *involves*

$$\mathcal{C}_R(m, n) = 2n^2(m - n/3) + O(mn)$$

operations. When the orthogonal basis $Q_1 = [q_1, \cdots, q_n] \in \mathbb{R}^{m \times n}$ of the range of A is sought, the basis is obtained by pre-multiplying the matrix $\begin{pmatrix} I_n \\ 0 \end{pmatrix}$ successively by $H^{(n)}$, $H^{(n-1)}, \cdots$, $H^{(1)}$. This process involves

$$\mathcal{C}_{Q_1}(m, n) = 2n^2(m - n/3) + O(mn)$$

arithmetic operations instead of

$$\mathcal{C}_Q(m, n) = 4(m^2 n - mn^2 + n^3/3) + O(mn)$$

operations when the whole matrix $Q = [q_1, \cdots, q_m] \in \mathbb{R}^{m \times m}$ must be assembled.

4.4.4 Givens Rotations

In the standard Givens process, the orthogonal matrices in (4.23) consist of $p = mn - n(n+1)/2$ plane rotations extended to the entire space $\mathbb{R}^m$. The triangularization of A is obtained by successively applying plane rotations $R^{(k)}$, so that

$$A^{(k+1)} = R^{(k)} A^{(k)} \text{ for } k = 1, \cdots, mn - n(n+1)/2 \text{ with } A^{(0)} = A.$$

Each of these rotations will be chosen to introduce a single zero below the diagonal of A: at step k, eliminating the entry $a_{i\ell}^{(k)}$ of $A^{(k)}$ where $1 \leq \ell \leq n$ and $1 < i \leq m$ can be done by the rotation

$$R_{ij,\ell}^{(k)} = \begin{pmatrix} 1 & & & & & & & & & \\ & \ddots & & & & & & & & \\ & & 1 & & & & & & & \\ & & & c & \cdot & \cdot & \cdot & -s & & \\ & & & \cdot & 1 & & & \cdot & & \\ & & & \cdot & & \ddots & & \cdot & & \\ & & & \cdot & & & 1 & \cdot & & \\ & & & s & \cdot & \cdot & \cdot & c & & \\ & & & & & & & & 1 & \\ & & & & & & & & & \ddots \\ & & & & & & & & & & 1 \end{pmatrix} \quad (4.37)$$

$$\uparrow i^{th} col. \qquad \uparrow j^{th} col.$$

where $1 \leq j \leq n$, and $c^2 + s^2 = 1$ are such that $\begin{pmatrix} c & -s \\ s & c \end{pmatrix} \begin{pmatrix} a_{j\ell}^{(k)} \\ a_{i\ell}^{(k)} \end{pmatrix} = \begin{pmatrix} a_{j\ell}^{(k+1)} \\ 0 \end{pmatrix}$. Two such rotations exist from which one is selected for numerical stability [33]. More precisely:

Theorem 4.9 *Let α, $\beta \in \mathbb{R}$ such that $\beta \neq 0$. There exist two rotations* $R = \begin{pmatrix} c & -s \\ s & c \end{pmatrix}$ *with $c = \cos\theta$ and $s = \sin\theta$ such that* $R\begin{pmatrix} \alpha \\ \beta \end{pmatrix} = \begin{pmatrix} \pm\rho \\ 0 \end{pmatrix}$ *with $\rho = \sqrt{\alpha^2 + \beta^2}$. One of them satisfies $\theta \in [-\frac{\pi}{2}, \frac{\pi}{2})$. It is determined by:*

$$\begin{bmatrix} \text{if } |\beta| > |\alpha|, & \tau = -\frac{\alpha}{\beta}, & s = \frac{1}{\sqrt{1+\tau^2}}, & c = s\tau, \\ \text{else} & \tau = -\frac{\beta}{\alpha}, & c = \frac{1}{\sqrt{1+\tau^2}}, & s = c\tau. \end{bmatrix} \qquad (4.38)$$

Proof. A rotation is a solution of the problem if and only if $s\alpha + c\beta = 0$. This condition is equivalent to $\cot\theta = -\frac{\alpha}{\beta}$. As soon as such a rotation is chosen, the first relation $c\alpha - s\beta = \pm\rho$ is automatically satisfied since a rotation is an isometry and therefore $\|R\begin{pmatrix} \alpha \\ \beta \end{pmatrix}\| = \rho$. There are two possible values for θ in the interval $[-\pi, \pi)$ and only one in the interval $[-\frac{\pi}{2}, \frac{\pi}{2})$. A simple calculation proves that this last solution is given by the procedure given in (4.38). ∎

The order of annihilation is quite flexible as long as applying a new rotation does not destroy a previously introduced zero. The simplest way is to organize column-wise eliminations. For instance by operating from the bottom of the columns and considering the notations of (4.37), the list of rotations are

$$R^{(1)}_{m-1\ m,1}, \ R^{(2)}_{m-2\ m-1,1}, \cdots, \ R^{(m-1)}_{1\ 2,1}, \ R^{(m)}_{m-1\ m,2}, \cdots$$

To illustrate the situation, let us consider a matrix $A \in \mathbb{R}^{4\times3}$. The previous order of elimination is

$$\begin{pmatrix} \cdot & \cdot & \cdot \\ 3 & \cdot & \cdot \\ 2 & 5 & \cdot \\ 1 & 4 & 6 \end{pmatrix}$$

where the order of the rotations is defined by the numbers.

To obtain the complete triangular factor R of (4.23), the total number of operations is $n^2(3m - n) + O(mn)$ which is roughly 1.5 times that required by the Householder's reduction. However, Givens rotations are of interest when the matrix A is of special structure, for example, Hessenberg or block structured.

4.5   Gram-Schmidt Orthogonalization

In its elementary form, the Gram-Schmidt process was introduced in Chapter 2, Section 2.2.3. Recall in that context that our objective was to compute an orthonormal basis $Q = [q_1, \cdots, q_n]$ out of the subspace spanned by the linearly independent column vectors of the matrix $A = [a_1, \cdots, a_n]$, in such a way that, for $1 \leq k \leq n$, the columns of $Q^{(k)} = [q_1, \cdots, q_k]$ is a basis of the subspace $\mathcal{A}_k$ spanned by the first k columns of A. The general procedure can be expressed by the following algorithm which structure is given by:

Algorithm 4.1 General Procedure for Gram-Schmidt Projections

```
Q(:,1)=A(:,1)/norm(A(:,1));
for   k = 1 : n-1
  % Pk orthogonal projection on span{Q(:,1),...,Q(:,k)},
   w = Pk(A(:,k+1));
   Q(:,k+1)=w/norm(w);
end
```

---

where P_k is the orthogonal projector onto the orthogonal complement of the subspace spanned by $\{a_1, \cdots, a_k\}$ (P_0 is the identity). $R = Q^\top A$ can be built step by step during the process.

Two different versions of the algorithm are obtained by expressing P_k in distinct ways:

CGS: For the Classical Gram-Schmidt (as in Algorithm 4.2), $P_k = I - Q_k Q_k^\top$, where $Q_k = [q_1, \cdots, q_k]$.

Unfortunately, this version has been proved by Björck [10] to be numerically unreliable except when it is applied two times which makes it equivalent to MGS.

MGS: To avoid the numerical instability of CGS, we use the Modified Gram-Schmidt. This method is based on the formula:

$$P_k = (I - q_k q_k^\top) \cdots (I - q_1 q_1^\top).$$

(See Chapter 2, Exercise 2.1). Numerically, using this alternative, the columns of Q are orthogonal up to a tolerance determined by the machine precision multiplied by the condition number of A [10]: MGS in Algorithm 4.3. By applying the algorithm a second time (i.e., with a complete re-orthogonalization), the columns of Q become orthogonal up to the tolerance determined by the machine precision parameter similar to the Householder or Givens reductions.

Algorithm 4.2 CGS: Classical Gram-Schmidt

```
function [Q,R]=cgs(A);
%
% Classical-Gram-Schmidt procedure
%
[n,m]=size(A);Q=zeros(n,m);R=zeros(m);
Q=A;R(1,1)=norm(A(:,1)) ;Q(:,1)=A(:,1)/R(1,1);
for k = 1:m-1,
    R(1:k,k+1)=Q(:,1:k)'*A(:,k+1);
    Q(:,k+1)=A(:,k+1)-Q(:,1:k)*R(1:k,k+1);
    R(k+1,k+1)=norm(Q(:,k+1)) ;Q(:,k+1) = Q(:,k+1)/R(k+1,k+1);
end;
```

Algorithm 4.3 MGS: Modified Gram-Schmidt

```
function [Q,R]=mgs(A);
%
% Modified-Gram-Schmidt procedure
%
[n,m]=size(A);Q=zeros(n,m);R=zeros(m);
R(1,1)=norm(A(:,1));Q(:,1)=A(:,1)/R(1,1);
for k = 1:m-1,
    w = A(:,k+1);
    for j = 1:k,
      R(j,k+1)=Q(:,j)'*w ;w = w - Q(:,j)*R(j,k+1) ;
    end ;
    R(k+1,k+1)=norm(w) ;Q(:,k+1) = w/R(k+1,k+1);
 end
```

The basic procedures involve $2mn^2 + O(mn)$ operations. CGS is based on BLAS-2 procedures but it must be applied two times, while MGS is based on BLAS-1 routines. By inverting the two loops of MGS, the procedure can proceed with a BLAS-2 routine. This is only possible when A is available explicitly (for instance, this is not possible with the Arnoldi procedure (See ...)). In order to proceed with BLAS-3 routines, a block version BGS is obtained by considering blocks of vectors instead of single vectors in MGS and by replacing the normalizing step by an application of MGS on the individual blocks, where blocks R_{ik} are square and of order p. The basic primitives used in that algorithm belong to BLAS-3 which insures full speed in the calculation on most of the computers. In [40], one proves that it is necessary to apply the normalizing step twice to reach the numerical accuracy of MGS.

Algorithm 4.4 BGS: Block Gram-Schmidt

```
function [Q,R] = bgs(A,p)
%BGS Block Gram Schmidt orthogonalization
%   This procedure is unreliable for ill conditioned systems
%
[n,m]=size(A); q=ceil(m/p);Q=A;s=min(p,m); K=1:s ;
R=zeros(m,m); [Q(:,K),R(K,K)]=mgs(Q(:,K)) ;
for k = 1:q-1,
    I=s+1:m ;
    R(K,I)=Q(:,K)'*Q(:,I);Q(:,I)=Q(:,I) - Q(:,K)*R(K,I) ;
    snew=min(s+p,m) ; K=s+1:snew;[Q(:,K),R(K,K)]=mgs(Q(:,K)) ;
    s=snew;
end
```

The new procedure becomes Algorithm 4.5.

Algorithm 4.5 B2GS: Safe Block Gram-Schmidt

```
function [Q,R] = b2gs(A,p)
%BGS  Safe block Gram Schmidt orthogonalization
%   This procedure is unreliable for ill conditioned systems
%   [Q,R] = b2gs(A,p)
[n,m]=size(A); q=ceil(m/p);Q=A;s=min(p,m);
K=1:s ;R=zeros(m,m); [Q(:,K),R(K,K)]=mgs(Q(:,K));
[Q(:,K),rho]=mgs(Q(:,K)) ;R(K,K)=rho*R(K,K) ;
for k = 1:q-1,
    I=s+1:m ;
    R(K,I)=Q(:,K)'*Q(:,I);Q(:,I)=Q(:,I) - Q(:,K)*R(K,I) ;
    snew=min(s+p,m) ; K=s+1:snew ;
    [Q(:,K),R(K,K)]=mgs(Q(:,K));[Q(:,K),rho]=mgs(Q(:,K)) ;
    R(K,K)=rho*R(K,K) ;s=snew;
end
```

It is easy to calculate that the number of operations involved by BGS is the same as with MGS ($2mn^2$ flops) and that B2GS involves $2mlp^2 = 2mnp$ additional flops to re-orthogonalize the blocks. That proves that the computational effort of B2GS equals to $(1 + \frac{1}{l})$ times the effort of MGS. In order to make small the price to pay, the number of blocks must be large. However, too small blocks do not bring the computational efficiency of BLAS-3. Therefore, there is a trade-off to find for each computer. A block size of 32 or 64 is often sufficient.

4.6   Least Squares Problem and QR Decomposition

For any orthogonal transform $Q \in \mathbb{R}^{m \times m}$, the problem (4.1) is equivalent to:

$$\|Q^T y - Q^T A z\|_2 = \min_{w \in \mathbb{R}^n} \|Q^T y - Q^T A w\| \qquad (4.39)$$

Let Q in (4.6) be the matrix obtained from the QR factorization $A = QR$, then Least squares problem reduces to:
Find $z \in \mathbb{R}^n$, such that:

$$\|c - Rz\|_2 = \min_{w \in \mathbb{R}^n} \|c - Rw\|, \text{ where } c = Q^T y. \qquad (4.40)$$

When $m > n$ and A is of full column rank, the Economy Version of the QR factorization (4.36) can be used and the matrix $\tilde{R}$ is non-singular. By denoting $c_i = Q_i^T b$, for $i = 1, 2$, the problem (4.40) has exactly one unique solution $w = \tilde{R}^{-1} c_1$. The corresponding residual is $r = y - Aw = Q_2 c_2 = Q_2 Q_2^T y = (I - Q_1 Q_1^T) y$ and $\|r\| = \|c_2\|$. Since $Q_2 Q_2^T = I - Q_1 Q_1^T$ is the orthogonal projection onto the subspace spanned by the column of Q_1 which is an orthonormal basis of $Im(A)$, the result illustrates the claims of Theorems 4.2 and 4.4.

Remark 4.2 *To get the problem (4.40), it is not even necessary to compute Q_1, since it is implicitly defined by the sequence of the Householder transformations: one can apply successively the Householder transforms $H^{(1)}, \cdots , H^{(s)}$ (where $s = \min(n, m - 1)$) on the left side of the matrix $[A, y]$ to get*

$$H^{(s)} H^{(s-1)} \cdots H^{(1)} [A, y] = \begin{pmatrix} R & c_1 \\ 0 & c_2 \end{pmatrix}, \qquad (4.41)$$

and then compute $x = R^{-1} c_1$.

Considering the numerical reliability of the solution obtained from Householder transformations, one can see that it relies on the quality of the solution of the linear system $\tilde{R}w = c_1$. Since A and $Q^T A$ have the same singular values $(\sigma_i)_{1,n}$, the condition number of the problem can be expressed by:

Proposition 4.4 *The condition number of the problem (4.1) is*

$$cond(A) = cond(\tilde{R}) = \frac{\sigma_1}{\sigma_n}. \qquad (4.42)$$

This clearly proves the superiority of the solution of (4.1) when obtained through a QR factorization rather than from the normal equations for which the condition number is squared as shown in (4.21).

4.7   Householder QR with Column Pivoting

In order to avoid computing the singular-value decomposition which is quite expensive for solving rank deficient linear least squares problems, the factorization (4.17) is often replaced by a similar factorization where S_r is not assumed to be diagonal. Such factorizations are called *complete orthogonal factorizations*. One way to compute such factorization relies on the procedure introduced by Businger and Golub [16]. It computes the QR factorization by Householder reductions with column pivoting. Factorization (4.35) is now replaced by

$$H^{(q)} \cdots H^{(2)} H^{(1)} A P_1 P_2 \cdots P_q = \begin{pmatrix} T \\ 0 \end{pmatrix}, \tag{4.43}$$

where matrices $P_1, \cdots, P_q$ are permutations and $T \in \mathbb{R}^{q \times n}$ is upper-trapezoidal. The permutations are determined so as to insure that the absolute values of the diagonal entries of T are non-increasing. To obtain a complete orthogonal factorization of A, the process performs first the factorization (4.43) and then compute an LQ factorization of T (equivalent to a QR factorization of T^T). The process ends up with a factorization similar to (4.17) with S_r being a non-singular lower-triangular matrix.

4.8   MATLAB Implementations

4.8.1   Use of the Backslash Operator

If A is an m by n matrix with $m < n$ or $m > n$ and y a column vector with m components, then x=A\y is the solution in the least squares sense to the under- or overdetermined system of equations $Ax = y$. The effective rank, r of A is determined from the QR decomposition with pivoting. A solution x is computed which has at most r non-zero components per column. If $r < n$, which is not the same solution as $A^+ y$.

4.8.2 QR Decompositions

1. [Q,R] = qr(A) where A is m by n, produces an m by n upper triangular matrix R and an m by m unitary matrix Q, so that $A = Q * R$.

2. [Q,R] = qr(A,0) where A is m by n, produces the "economy size" decomposition, so that if $m > n$, only the first n columns of Q and the first n rows of R are computed. If $m \leq n$, the result is the same as [Q,R] = qr(A).

3. [Q,R,P] = qr(A) where A is m by n, produces an m by n upper triangular matrix R and an m by m unitary matrix Q and a permutation matrix P so that $A * P = Q * R$. The column permutation matrix P is chosen so that the diagonal elements of R are decreasing:

$$R(1,1) \geq R(2,2) \geq ...R(r,r)$$

4. [Q,R,P] = qr(A,0) produces an "economy size" decomposition in which P is a permutation vector, so that A(:,P) = Q*R.

4.9   Exercises

Exercise 4.1 *In the case of (4.2):*
1. Assume $z = (z_1, z_2)$. *Show that the function* $\phi(z)$ *(4.7) is given by the quadratic form:*

$$\phi(z) = \frac{1}{2}(b_{11}z_1^2 + b_{22}z_2^2) + b_{12}z_1z_2 - c_1z_1 - c_2z_2,$$

then prove that the minimum of $\phi(z)$ *is reached when* $z = w = (w_1, w_2)$ *with* $Bw = c$.
2. Give the solution $y = w_1x + w_2$ *in terms of* $b_{ij} : 1 \le i, j \le 2$ *and* c_1, c_2.

Exercise 4.2 *Repeat Exercise 1 in the case of (4.3) with* $z = (z_1, z_2, z_3)$ *and* $w = (w_1, w_2, w_3)$.

Exercise 4.3

Write a segment of `MATLAB` program that generates the pseudo-inverse of a rectangular matrix $A \in \mathbb{R}^{m \times n}$, $m < n$, $m = n$, $m > n$ using the `svd` decomposition command.

Exercise 4.4

Write a segment of `MATLAB` program that solves the linear least squares problem $x = A^+y$ for $A \in \mathbb{R}^{m \times n}$ from a QR-factorization. The matrix is supposed to be of full column rank when $m \ge n$ or full row rank when $m < n$. In the latter case (undetermined linear system), consider the QR factorization of A^T. Determine in each situation, the number of arithmetic operations involved.

Exercise 4.5

By using adequate Givens rotations to perform the QR factorization of an upper Hessenberg matrix (see Definition 5.3), prove that the orthogonal factor Q is upper Hessenberg as well.

4.10   Computer Exercises

Computer Exercise 4.1 Testing Orthogonality

1. Consider the following MATLAB program:

```
m=100; n=10; kmax=15;
A1=ones(m,n)-2*rand(m,n);
[U,S,V]=svd(A1);
A0=A1-S(n,n)*U(:,n)*V(:,n)';
C0=cond(A0),
C=[]; OrthMGS=[];OrthHouse=[];
for k=1:kmax,
     tau=10^(-k);
     B=(1-tau)*A0+tau*A1;
     C=[C,cond(B)];
     Q=mymgs(B); % Q=mymgs(Q);
     OrthMGS=[OrthMGS,norm(eye(n)-Q'*Q)];
     [Q,R]=qr(B,0);
     OrthHouse=[OrthHouse,norm(eye(n)-Q'*Q)];
end
semilogy([1:k],C,'kx-',[1:k],OrthMGS,'k+-',[1:k],OrthHouse,'ko-');
legend('Condition number','Orthog. MGS','Orthog. Householder');
```

in which the function `mymgs` implements the Modified Gram Schmidt procedure as done by Algorithm 4.3. Which method is implemented by the `qr` function of MATLAB?

Let $\sigma_1 \geq \sigma_2 \geq \cdots \geq \sigma_n$ be the singular values of A1. What are the singular values of A1? What are the singular values of B for each k?

2. Run the program and comment on the results. Confirm that it can be observed that

$$\|I - Q_{mgs}^T Q_{mgs}\| = O(Cond(B)\ \epsilon) \text{ and } \|I - Q_{house}^T Q_{house}\| = O(\epsilon),$$

where Q_{mgs} and Q_{house} are, respectively, the results of the orthogonalization by `mgs` and `qr`, and where ϵ is the epsilon machine precision parameter.

3. In the program, replace line

$$Q=mymgs(B);$$

by

$$Q=mymgs(B); \ Q=mymgs(Q);$$

Rerun the program and comment.

Computer Exercise 4.2 Global Positioning System

The Global Positioning System (GPS) is used in everyday life. It allows to get the geographical position at any time and any location of the world through a receiver decoding signals from several satellites. For a constellation of 24 satellites, any point of the earth is always in view of 6 to 12 satellites.

Let $r = (\alpha, \beta, \gamma)^T$ be the three coordinates of the receiver; they are the unknowns of the problem. Their determination is based on the knowledge of the travelling times of the signals from the satellites to the receiver. The satellites embed highly precise clocks (atomic clocks) but the receiver time is only based on a clock with lower precision. Therefore, it is necessary to evaluate the time bias between the local clock and the precise time and this bias Δt becomes an unknown as well. In order to have the same unit for all the unknowns the fourth unknown is chosen to be the distance $c \, \Delta t$ where c is the speed of light.

Let us assume that the receiver gets signals from N satellites S_i $(i = 1, \cdots, N)$ and that the coordinates of satellite S_i are $s_i = (\alpha_i, \beta_i, \gamma_i)^T$ [1]. These coordinates are sent by satellite S_i to the receiver with the date t_{e_i} of the emission of the signal. The receiver gets this signal at the biased time t_{r_i}.

1. **Mathematical Modeling:** For $i = 1, \cdots, N$, prove the following equation

$$\|r - s_i\| + c \, \Delta t \;=\; c \, (t_{r_i} - t_{e_i}). \tag{4.44}$$

By definition, let:

- $\rho_i = c \, (t_{r_i} - t_{e_i})$, for $i = 1, \cdots, N$;
- $a_i = (\alpha_i, \beta_i, \gamma_i, \rho_i)^T \in \mathbb{R}^4$, for $i = 1, \cdots, N$;
- $x = (\alpha, \beta, \gamma, c \, \Delta t)^T \in \mathbb{R}^4$.

How many satellites are a priori needed to have enough information to solve the problem?

2. **Problem solution [8].** Prove that for $i = 1, \cdots, N$, equation (4.44) is equivalent to

$$\frac{1}{2} < a_i, a_i > - < a_i, x > + \frac{1}{2} < x, x > \;=\; 0, \tag{4.45}$$

where $< ., . >$ is the Minkowski functional defined by

$$< u, v >= \mu_1 \nu_1 + \mu_2 \nu_2 + \mu_3 \nu_3 - \mu_4 \nu_4, \tag{4.46}$$

for any vectors $u = (\mu_1, \mu_2, \mu_3, \mu_4)^T$ and $v = (\nu_1, \nu_2, \nu_3, \nu_4)^T$. Is the Minkowski functional an inner-product?

Prove that the system to solve can be expressed under a matrix form by

$$Ax = a + \lambda e, \tag{4.47}$$

---

[1] The chosen coordinate system may be given by the center of the earth, the equatorial plane and the rotation axis.

where

- $a = \frac{1}{2}(<a_1, a_1>, \cdots, <a_N, a_N>)^T$;

- $A = \begin{pmatrix} \alpha_1 & \beta_1 & \gamma_1 & -\rho_1 \\ \alpha_2 & \beta_2 & \gamma_2 & -\rho_2 \\ \vdots & \vdots & \vdots & \vdots \\ \alpha_N & \beta_N & \gamma_N & -\rho_N \end{pmatrix}$;

- $\lambda = \frac{1}{2} <x, x>$ and $e = (1, \cdots, 1)^T \in \mathbb{R}^{N \times 4}$.

From (4.47), the solution x is chosen to satisfy $x = A^+(a + \lambda e)$ where A^+ is the pseudo-inverse of A. Then, prove that the solution is given by the quadratic equation

$$\lambda^2 <A^+e, A^+e> \; + \; 2\lambda \left(<A^+a, A^+e> -1\right) + \; <A^+a, A^+a> = 0. \quad (4.48)$$

How does one select the correct solution?

3. Write the MATLAB function:

```
function [r, Deltat] = GPS(S,Te,Tr)
% Input
%   S  : (N x 3)-matrix : the i-th row contains the coordinates
%                                     of Satellite Si.
%   Te : N-vector : the i-th entry is the emitting date of the
%                               signal received from Si.
%   Tr : N-vector : the i-th entry is the receiving date of the
%                               signal received from Si.
% Output
%   r  : 3-vector containing the coordinates of the receiver.
%   Deltat : bias of the receiver clock such that
%                   (Tr(i) - Deltat) is the precise receiving time,
%                   for any i.
  ...
end
```

Chapter 5

Algorithms for the Eigenvalue Problem

5.1 Basic Principles

5.1.1 Why Compute the Eigenvalues of a Square Matrix?

The eigenvalues of an $n \times n$ square matrix are canonical parameters of the corresponding linear application $\mathcal{L}$ on a finite dimensional vector space E of dimension n.

Given the application $\mathcal{L}$, the matrix A depends on a reference basis $\mathcal{B}$ in the vector space E, which we note by $A = M_{\mathcal{B}}(\mathcal{L})$. When changing the reference basis in E, the matrix A is changed into a matrix $\tilde{A}$ that is similar to A, i.e., $\tilde{A} = X^{-1}AX$, where the columns of $X = [x_1, \cdots, x_n]$ are the vectors of the new basis expressed in terms of the previous one. A natural inquiry consists in finding a new basis, such that $\tilde{A}$ has the simplest form, for example, $\tilde{A}$ becomes a diagonal matrix. In that case, one "decouples" the linear application $\mathcal{L}$ into n "scalar" applications.

$$Ax_i = \lambda_i x_i, \text{ for } i = 1, \cdots, n. \tag{5.1}$$

This leads to the set values $\Lambda(A) = \{\lambda_1, \cdots, \lambda_n\}$, such that the matrix $(A - \lambda_i I)$ is singular. The eigenvalues of A are the roots of the polynomial $p_A(\lambda) = \det(A - \lambda I)$. This polynomial is called the characteristic polynomial; it is independent from the basis being considered (i.e., for any basis X, $\det(X^{-1}AX - \lambda I) = \det(X^{-1}(A - \lambda I)X) = \det(A - \lambda I)$). For the sake of illustration, we look now at two situations that use eigenvalues.

First Illustration: Linear Recurrence

Let us assume that some timely phenomenon is described by expressing $x_i(k)$ the number of persons of a population that lies in class i at time k. The phenomenon is therefore described by the sequence $x(0), x(1), \cdots, x(k), \cdots$ We assume in addition that passing from time k to time $k+1$ is governed by a matrix multiplication model:

$$x(k+1) = Ax(k),$$

We can thus write $x(k) = A^k x(0)$, where $x(0)$ is the class repartition. If A is diagonal, then it is easy to compute any power of A. Otherwise, if A is similar to a diagonal matrix D, i.e., $A = PDP^{-1}$, with $P \in \mathbb{R}^{n \times n}$ a change of basis (invertible) matrix and

$$D = \mathrm{diag}(\lambda_1, \cdots, \lambda_n).$$

This leads to:

$$A^k = PD^kP^{-1},$$

and one concludes, that $x_i(k)$ is a linear combination of the k powers of the eigenvalues $\lambda_1, \cdots, \lambda_n$, for $i = 1, \cdots, n$. Thus, if all eigenvalues have absolute values less than 1, then for any $i = 1, \cdots, n$, $\lim_{k \to \infty} x_i(k) = 0$. Otherwise, if at least one eigenvalue of A has its absolute value greater than 1, the phenomenon will have an "explosive" behavior, independently of $x(0)$.
For example, let us consider the Fibonacci sequence, defined by the recurrence:

$$\alpha_0 = 1, \ \alpha_1 = 1, \ \text{and}$$

$$\alpha_{k+1} = \alpha_k + \alpha_{k-1}, \ \text{for } k \geq 1.$$

It can be rewritten in the form of a linear recurrence, as follows:
$x(k) \equiv \begin{pmatrix} \alpha_{k+1} \\ \alpha_k \end{pmatrix} = \begin{pmatrix} 1 & 1 \\ 1 & 0 \end{pmatrix} x(k-1)$. The eigenvalues of the matrix are
$\lambda_1 = \frac{1+\sqrt{5}}{2}$ and $\lambda_2 = \frac{1-\sqrt{5}}{2}$.
One concludes that $\alpha_k = \frac{\sqrt{5}}{5} \left(\lambda_1{}^k - \lambda_2{}^k \right)$.
This is the case of a linear recurrence that goes to infinity when $k \to \infty$.

Second Illustration: Stability of Systems of Differential Equations

In this second example, we consider a continuous time dependent process, described by a homogeneous linear system of ordinary differential equations:

$$\begin{cases} \frac{dx}{dt} = Ax(t) \text{ for } t \geq 0, \\ x(0) = x_0 \end{cases}$$

where $x(t) \in \mathbb{R}^n$.
A classical question is to know whether the system is asymptotically stable: is the solution $x(t)$ converging to 0 when t goes to infinity? Let us assume that the matrix A is similar to a diagonal matrix. By an approach similar to the one considered in the previous example with linear recurrences, it can be proved that the components of $x(t)$ are fixed linear combinations of $e^{\lambda_i t}$ where the eigenvalues (possibly complex) of the matrix A are $\{\lambda_i, \ i = 1, \cdots, n\}$. The system is stable when all the eigenvalues are with a negative real part.

5.1.2   Spectral Decomposition of a Matrix

Let us now complete the definition of the problem. The definition and the Schur decomposition of a matrix have been introduced in Section 2.2.6. For the sake of completeness, we remind the readers the main results with references to the previous proofs when necessary. This chapter intends to provide effective algorithms that solve eigenvalue problems.

Throughout the chapter, the set $\mathbb{K}$ is either the set $\mathbb{C}$ of complex numbers or the set $\mathbb{R}$ of real numbers. Let us remember some definitions already introduced in Chapter 2.

Definition 5.1 *Let A be a square matrix of order n: $A \in \mathbb{K}^{n \times n}$. The characteristic polynomial of A is the polynomial defined by $p_A(\lambda) = \det(A - \lambda I)$. The eigenvalues of A are the roots of p_A and their set $\Lambda(A)$ is the spectrum of A. There are always n eigenvalues, possibly complex and possibly multiple. In the class of matrices which are similar to A, all the matrices have the same characteristic polynomial.*

A vector $x \in \mathbb{K}^n$ is an eigenvector corresponding to the eigenvalue $\lambda \in \mathbb{K}$ if and only if this is a non-zero vector which satisfies $Ax = \lambda x$. The null space $\text{Null}(A - \lambda I)$ is the subspace of all the eigenvectors corresponding to the eigenvalue λ with the null vector appended to the set. This set is an invariant set for A and is called eigenspace of A corresponding to λ.

The matrix A is said to be diagonalizable on $\mathbb{K}$ if there exists a basis $X = [x_1, \cdots, x_n] \in \mathbb{K}^{n \times n}$ such that $D = X^{-1} A X$ is a diagonal matrix.

If λ is an eigenvalue of A then there exists at least one corresponding eigenvector, or equivalently, the dimension of the eigenspace $\text{Null}(A - \lambda I)$ is not smaller than 1. This motivates the following proposition.

Proposition 5.1 *If the n eigenvalues of A are all distinct in $\mathbb{K}$, then A is diagonalizable in $\mathbb{K}$.*

Proof.   When the n eigenvalues are distinct, the whole space $\mathbb{K}$ can be expressed by the direct sum: $\mathbb{K} = \bigoplus_{i=1}^{n} \text{Null}(A - \lambda_i I)$. This proves that all the eigenspaces are of dimension 1. By selecting n non-zero vectors $x_1, \cdots, x_n$ as columns of $X = [x_1, \cdots, x_n]$ in these eigenspaces, the basis X is such that $AX = XD$ where $D = \text{diag}(\lambda_1, \cdots, \lambda_n)$ and therefore $X^{-1} A X = D$.   ∎

Remark 5.1 *There exist non-diagonalizable matrices. First, a real matrix may have complex eigenvalues and in this case it is not $\mathbb{R}$-diagonalizable. But even with complex arithmetic, some matrices are not diagonalizable; therefore, from Proposition 5.1, they must have multiple eigenvalues. For instance, the matrix $A = \begin{pmatrix} 0 & 1 \\ 0 & 0 \end{pmatrix}$ is not diagonalizable neither in $\mathbb{R}$ nor in $\mathbb{C}$.*

The next theorem expresses the fundamental result. Some algorithms follow the proof. The superscript $*$ on a matrix denotes its adjoint matrix (i.e., its conjugate transpose).

Theorem 5.1 (Complex Schur Factorization) *Let $A \in \mathbb{C}$ be a complex matrix. It is similar to an upper-triangular matrix: there exists a unitary matrix $U \in \mathbb{C}^{n \times n}$ ($U^*U = I$ and therefore $U^* = U^{-1}$) such that the matrix $T = U^*AU$ is upper-triangular. The factorization is therefore expressed by:*

$$A = UTU^*. \tag{5.2}$$

Proof. This is proved by induction on the matrix order n. See the proof of Theorem 2.4. ∎

Remark 5.2 *Since the determinant of a triangular matrix is equal to the product of all its diagonal entries, we may express the characteristic polynomial as $p_A(\lambda) = \det(A - \lambda I) = \det(T - \lambda I) = \prod_{i=1}^{n}(T_{ii} - \lambda)$. This proves that the diagonal entries of T are the eigenvalues of A.*
The Schur factorization is not unique since the order of appearance of the eigenvalues in the proof may follow any order.

Proposition 5.2 (Shift-and-Invert) *Let $A \in \mathbb{K}^{n \times n}$ ($\mathbb{K} = \mathbb{R}$ or $\mathbb{C}$) with the spectrum $\Lambda(A) = \{\lambda_1, \cdots, \lambda_n\}$. For any $\mu \in \mathbb{K}$ which is not an eigenvalue of A, the spectra of $(A - \mu I)$ and $(A - \mu I)^{-1}$ are:*

$$\Lambda(A - \mu I) = \{\lambda_1 - \mu, \cdots, \lambda_n - \mu\}, \tag{5.3}$$

$$\Lambda\left((A - \mu I)^{-1}\right) = \{\frac{1}{\lambda_1 - \mu}, \cdots, \frac{1}{\lambda_n - \mu}\}. \tag{5.4}$$

Proof. Let us consider the Schur factorization (5.2). Therefore $A - \mu I = U(T - \mu I)U^*$ and $(A - \mu I)^{-1} = U(T - \mu I)^{-1}U^*$ are the Schur factorizations of the two considered matrices. ∎

The next theorem expresses the highest reduction that can be considered for any general matrix.

Theorem 5.2 (Jordan Decomposition) *If $A \in \mathbb{C}^{n \times n}$, then there exists a non-singular $X \in \mathbb{C}^{n \times n}$ such that $J = X^{-1}AX = \mathrm{diag}(J_1, \cdots, J_t)$ where*

$$J_i = \begin{pmatrix} \lambda_i & 1 & & \cdots & 0 \\ 0 & \lambda_i & \ddots & & \vdots \\ & & \ddots & \ddots & \ddots \\ \vdots & & & \ddots & \ddots & 1 \\ 0 & \cdots & & & 0 & \lambda_i \end{pmatrix} \in \mathbb{C}^{m_i \times m_i},$$

and $m_1 + \cdots + m_t = n$. The blocks J_i are referred as Jordan blocks.

Proof. See, for example, reference [58]. ∎

When $m_i = 1$, the block J_i is reduced to the eigenvalue λ_i. When $m_i > 1$, the Jordan block J_i is associated with a multiple eigenvalue λ_i. It must be noticed that several Jordan blocks can be associated with the same eigenvalue. The algebraic multiplicity of an eigenvalue (i.e., the multiplicity as root of the characteristic polynomial) is therefore equal to the sum of the orders of all the Jordan blocks associated with this eigenvalue.

Proposition 5.3 *The eigenspace of the Jordan block J_i is of dimension 1:* $\dim Null(J_i - \lambda_i I_{m_i}) = 1$.

Proof. Left as an exercise (see Exercise 5.1). ∎

Remark 5.3 *There is a relation between the Schur and the Jordan decompositions. Let $A = XJX^{-1}$ be the Jordan decomposition and Q and R obtained from the QR-factorization of X: $X = QR$ where Q is a unitary matrix and R is an upper-triangular matrix. The matrix R is non-singular since X is non-singular. Therefore, $A = Q(RJR^{-1})Q^*$. Since the matrix $T = RJR^{-1}$ is upper triangular as a result of the multiplication of three upper-triangular matrices, the Schur factorization is obtained.*

Remark 5.4 (Non-stability of the Jordan Decomposition) *The Jordan form appears to be the finest decomposition of a matrix. However, it cannot be easily determined through numerical computations because the decomposition is not continuous with respect to the matrix: for any matrix $A \in \mathbb{K}^{n \times n}$ and for any $\epsilon > 0$, there exists $B \in \mathbb{K}^{n \times n}$ such that $\|A - B\| \leq \epsilon$ and the n eigenvalues of B are all distinct. In other words, the set of the diagonalizable matrices is dense in the set of matrices: the determination of the Jordan decomposition is an ill-posed problem (see Exercise 5.2). The Schur factorization is the decomposition on which the algorithms mainly rely.*

Corollary 5.1 (Expression of the Characteristic Polynomial) *The coefficients of the characteristic polynomial $p_A(\lambda) = \det(A - \lambda I)$ are:*

$$p_A(\lambda) = (-1)^n \lambda^n + (-1)^{n-1}\gamma_1 \lambda^{n-1} + (-1)^{n-2}\gamma_2 \lambda^{n-2} + \cdots - \gamma_{n-1}\lambda + \gamma_n,$$

where (γ_i) are the elementary symmetric polynomials in the eigenvalues of A. In particular:

$$\gamma_1 = \sum_{i=1}^{n} \lambda_i = \mathrm{trace}(A), \quad and \tag{5.5}$$

$$\gamma_n = \prod_{i=1}^{n} \lambda_i = \det(A). \tag{5.6}$$

Proof. The relations (5.5) and (5.6) are easy to prove. The other coefficients can be obtained from the Newton identities (see [35], pp. 166-168). ∎

Corollary 5.2 (Hermitian Matrices) *A Hermitian matrix $A \in \mathbb{C}^{n \times n}$ is unitarily similar to a real diagonal matrix: There exists a unitary matrix $U \in \mathbb{C}^{n \times n}$ ($U^*U = I$) such that the matrix $D = U^*AU$ is real and diagonal. When A is a real symmetric matrix, the matrix U is real and orthogonal ($U^TU = I$).*

Proof. Let us consider the situation when A is Hermitian: $A^* = A$. Its Schur factorization $T = U^*AU$ satisfies: $T^* = (U^*AU)^* = U^*AU = T$. Consequently, the matrix T is Hermitian and upper-triangular; the only possibility of such a matrix is to be a diagonal matrix. Moreover, its diagonal entries are equal to their conjugate and this proves that they are real. ■

Theorem 5.3 (Real Schur Factorization) *Let $A \in \mathbb{R}^{n \times n}$. It is orthogonally similar to a quasi-upper triangular matrix: there exists an orthogonal matrix $Q \in \mathbb{R}^{n \times n}$ such that the matrix $T = Q^T A Q$ is block-upper triangular, with diagonal blocks of order 1 or 2. The 2-dimensional blocks correspond to pairs of conjugate eigenvalues.*

Proof. The proof of Theorem 5.1 must be adapted to the real computation. The induction can progress one or two units depending on the following cases: (1) when $\lambda \in \mathbb{R}$, and (2) when λ is not real and therefore combined with its conjugate $\bar{\lambda}$. ■

There exist inequalities on the modulus of the eigenvalues which restrict their localization in the complex plane. The easiest one is given by any matrix norm which is induced from a vector norm.

Proposition 5.4 *For any matrix norm of $\mathbb{C}^{n \times n}$ which is induced by a vector norm of $\mathbb{C}^n$, the following inequality stands:*

$$\rho(A) \leq \|A\|,$$

where $\rho(A) = \max\{|\lambda| \,|\, \lambda \in \Lambda(A)\}$ is the spectral radius of A.

Proof. Let λ be an eigenvalue of maximum modulus and x a corresponding eigenvector and let consider some vector norm and its induced matrix norm. The result follows from:

$$\rho(A) = |\lambda| = \frac{\|Ax\|}{\|x\|} \leq \|A\|.$$

■

In consequence, the spectrum $\Lambda(A)$ is included in the disk centered at the origin and of radius $\|A\|$. The following theorem provides a finer localization.

Theorem 5.4 (Gershgorin) *Let $A = (a_{ij}) \in \mathbb{C}^{n \times n}$, the spectrum of A is included in the set $(\cup_{i=1}^n \mathcal{D}_i)$ where $\mathcal{D}_i$ is the closed disk in $\mathbb{C}$ centered at a_{ii} and of radius $\sum_{j \neq i} |a_{ij}|$.*

Proof. Let λ be an eigenvalue of A. Let us assume that λ is not a diagonal entry of A (otherwise it obviously belongs to the corresponding disk $\mathcal{D}_i$). The matrix A can be split into $A = D + E$ such that D is the diagonal matrix defined by $\text{diag}(D) = \text{diag}(A)$. From the initial assumptions, the matrix $D - \lambda I + E$ is singular whereas the matrix $(D - \lambda I)$ is non-singular. From the decomposition

$$A - \lambda I = (D - \lambda I)[I + (D - \lambda I)^{-1}E]$$

it appears that the matrix $[I + (D - \lambda I)^{-1}E]$ is necessarily singular. This implies that $\|(D - \lambda I)^{-1}E\|_\infty \geq 1$, otherwise the matrix $[I + (D - \lambda I)^{-1}E]$ would be non-singular and its inverse would be expressed by a series of $H = (D - \lambda I)^{-1}E$. Therefore there exists an index i of a row of the matrix $(D - \lambda I)^{-1}E$ such that

$$\sum_{j \neq i} \frac{|a_{ij}|}{|a_{ii} - \lambda|} \geq 1,$$

which proves that $\lambda \in \mathcal{D}_i$. ∎

Theorem 5.5 (Cayley-Hamilton) *The characteristic polynomial p_A of the matrix $A \in \mathbb{C}^{n \times n}$ satisfies $p(A) = 0$.*

Proof. See reference [32]. ∎

In the ring of polynomials, the set of all the polynomials $q \in \mathbb{C}[X]$ such that $q(A) = 0$ is an ideal of $\mathbb{C}[X]$ which includes the characteristic polynomial p_A.

Definition 5.2 (Minimal Polynomial) *The minimal polynomial is the monic polynomial q_A that spans the ideal*

$$\mathcal{I} = \{q \in \mathbb{C}[X] \mid q(A) = 0\}.$$

The characteristic polynomial is a multiple of the minimal polynomial.

Remark 5.5 *It can be shown that the exponent of the factor $(\lambda - \mu)$ in the factorized expression of the minimal polynomial is the largest order of the Jordan blocks associated with the eigenvalue μ.*

The last mathematical results of this section concern fundamental properties of the symmetric matrices.

Theorem 5.6 (Symmetric Perturbation) *Let two symmetric matrices $A \in \mathbb{R}^{n \times n}$ and $\Delta \in \mathbb{R}^{n \times n}$. Therefore, for every eigenvalue $\tilde{\lambda} \in \mathbb{R}$ of the matrix $A + \Delta$, there exists an eigenvalue $\lambda \in \mathbb{R}$ of the matrix A, such that*

$$|\tilde{\lambda} - \lambda| \leq \|\Delta\|_2. \tag{5.7}$$

Proof. See reference [64]. ∎

Theorem 5.7 (Interlacing Property) *Let $B \in \mathbb{R}^{(n-1) \times (n-1)}$ the principal submatrix of order $n - 1$ of a symmetric matrix $A \in \mathbb{R}^{n \times n}$ (i.e., $A = \begin{pmatrix} B & a \\ a^T & \alpha \end{pmatrix}$ where $a = A_{1:n-1,n}$ and $\alpha = A_{nn}$). By numbering in ascending order the eigenvalues $\lambda_1 \leq \cdots \leq \lambda_n$ of A and the eigenvalues $\mu_1 \leq \cdots \leq \mu_{n-1}$ of B, the following inequalities stand:*

$$\lambda_1 \leq \mu_1 \leq \lambda_2 \leq \mu_2 \leq \cdots \leq \mu_{n-1} \leq \lambda_n.$$

Proof. See reference [64]. ∎

5.1.3   The Power Method and its By-Products

To make easier the presentation, we restrict the situation to a real matrix, but it can be easily extended to the complex scene. Let us assume that the eigenvalue λ of largest modulus of the matrix $A \in \mathbb{R}^{n \times n}$ is unique; therefore, it is real and $\lambda = \pm \rho(A)$. Algorithm 5.1 defines a sequence of real numbers $\{\lambda_k | k = 0, 1, ...\}$ that converge to λ and a sequence of vectors $\{x^{(k)} | k = 0, 1, ...\}$ that approximate the eigenvector corresponding to λ. Specifically for $x^{(0)}$ arbitrary with $||x^{(0)}|| = 1$:

$$\text{for } k = 0, 1, 2, ..., \ \lambda_k = (x^{(k)})^T A x^{(k)}, \ x^{(k+1)} = (\frac{1}{||Ax^{(k)}||}) A x^{(k)}.$$

Note that in this version, the power method algorithm is expressed under a simplified expression. For a library version, it would be necessary to control the number of iterations and to define the convergence tolerance parameter tol proportionaly to $|\lambda|$ or $\|A\|$.

Algorithm 5.1 Algorithm of the Power Method

```
function [lambda,x]=mypower(A,tol)
n=length(A);k=0;x=rand(n,1);x = x/norm(x);
nr=1;
while nr > tol
    k=k+1;y=A*x; lambda= x'*y;
    r= y - lambda*x;nr=norm(r);
    x = y/norm(y);
end
```

Theorem 5.8 *Let $\lambda^{(k)}$ and $x^{(k)}$ be the value of* lambda *and of* x *at the kth ($k \geq 1$) iteration of Algorithm 5.1. The sequence $(\lambda^{(k)})_{k\geq 1}$ converges to λ the eigenvalue of largest modulus for almost any initial vector x_0. Any accumulation point x of the sequence $(x^{(k)})_{k\geq 1}$ is an eigenvector of A corresponding to λ. The convergence of the sequence $(\lambda^{(k)})_{k\geq 1}$ is linear with a rate of convergence equal to $|\frac{\lambda}{\lambda_2}|$ where λ_2 is the second eigenvalue of A of largest modulus (by assumption $|\lambda_2| < \rho(A)$).*

Proof. We prove the theorem in the situation of a symmetric matrix but the proof is almost the same for a non-symmetric matrix. Let us assume that $A = QDQ^T$ where Q is an orthogonal matrix of eigenvectors of A and $D = \text{diag}(\lambda, \lambda_2, \cdots, \lambda_n)$ the diagonal matrix with the eigenvalues of A as diagonal entries. It is easy to see that the iterated vector can be expressed by $x^{(k)} = \alpha_k A^k x_0$ where α_k is some real that insures that $||x^{(k)}|| = 1$. By introducing the vector $u = Q^T x_0$, we get $x^{(k)} = \alpha_k Q D^k u$ and therefore

$$x^{(k)} = \lambda^k \alpha_k Q \begin{pmatrix} \mu_1 \\ (\frac{\lambda_2}{\lambda})^k \mu_2 \\ \vdots \\ (\frac{\lambda_n}{\lambda})^k \mu_n \end{pmatrix}$$

where μ_i $(i = 1, \cdots, n)$ are the components of u. The direction of the vector $x^{(k)}$ converges to the direction of q_1, first column of Q if $\mu_1 \neq 0$. The property $\mu_1 = 0$ corresponds to the special situation where the initial vector is exactly orthogonal to the sought eigenvector. When $\lambda < 0$, the sequence $x^{(k)}$ alternates at the limit. ∎

This algorithm is very popular for its simplicity. Moreover, it does not involve transformations of the matrix A which is only used through its multiplication with vectors. This property is of special interest for large and sparse matrices. However, the main drawback of the method is its very slow convergence when the ratio $|\frac{\lambda_2}{\lambda}|$ is almost equal to 1 which is a common situation for large matrices.

The Inverse Iteration Method

When an eigenvalue λ is already approximated by some close value μ, the Power method can be adapted to compute the corresponding eigenvector. Let us assume that $\epsilon = |\lambda - \mu|$ satisfies $\epsilon \ll |\lambda_2 - \mu|$ where λ_2 is the closest eigenvalue of A after λ. Following the result of Proposition 5.2, the Power method can be applied to the matrix $C = (A - \mu I)^{-1}$, resulting in a small convergence rate which means a fast convergence: $|\frac{\epsilon}{|\lambda_2 - \mu|}| \ll 1$. By noticing that $y = Cx$ implies that the vector $\frac{x}{\|y\|} = (A - \mu I)\frac{y}{\|y\|}$ is the residual associated to the pair $(\mu, \frac{y}{\|y\|})$. The norm of the residual is therefore $\frac{1}{\|y\|}$ and it can be used to control the convergence. The corresponding Algorithm 5.2 is called the Inverse-Power Method.

Algorithm 5.2 Algorithm of the Inverse Iteration Method

```
function [lambda,x]=inverseiter(A,mu,tol)
n=length(A);B=A-mu*eye(n);
%and LU-factorization
x=rand(n,1);x = x/norm(x);alpha=1;
while alpha > tol
    y=B \ x ; % using the LU-factors
    alpha=norm(y) ;x=y/alpha;alpha=1/alpha;
end
lambda=x'*(A*x) ;
end
```

The method involves a linear system solution at every iteration but only one LU-factorization is performed at the beginning since the matrix B is the same for all iterations.

Algorithm 5.2 is included in the libraries to determine an eigenvector of an eigenvalue μ already computed at the machine precision accuracy. One may be surprised by this approach since the matrix $A - \mu I$ is numerically singular and the error on the solution of the system $Bx = y$ is large. Fortunately, the error vector is also mainly in the direction of the sought eigenvector (see, for example, reference [55]) and after the normalizing step the approximation is good. Except for very close eigenvalues, only one iteration is usually sufficient to get the correct accuracy.

Rayleigh Quotient Iteration

The Inverse Iteration Method can also be adapted into a new algorithm in which the shift parameter is updated at each iteration. The corresponding method is the Rayleigh Quotient Iteration method which is expressed by Algorithm 5.3: at every iteration, the shift λ is estimated by the Rayleigh Quotient of the current iterate x: $\lambda = \rho(x) = x^T A x$. This makes the convergence rate factor smaller at each iteration and therefore it speeds up the convergence. There is a price to pay since the method implies an LU-factorization of the matrix $(A - \lambda I)$ at each iteration.

Algorithm 5.3 Algorithm for the Rayleigh Quotient Iteration

```
function [lambda,x]=quotient_rayleigh(A,mu,tol)
 x=rand(n,1) ; x = x/norm(x) ;
 y=(A - mu*eye(n)) \ x ; alpha = norm(y) ;
 while 1/alpha >=  tol,
    x = y / alpha;lambda = x'*(A*x) ;
    y = (A - lambda*eye(n))\ x ;
    alpha = norm(y) ;
  end
```

When the matrix A is real and symmetric, the Rayleigh Quotient Iteration converges cubically. When the matrix is non-symmetric, the process converges only quadratically (for the proof, see reference [55]).

5.2 QR Method for a Non-Symmetric Matrix

In this section, it is assumed that the matrix A is real and that computations with complex numbers will be delayed as long as possible. When the
matrix is non-symmetric, some eigenvalues can be complex and therefore at
least the final step of the algorithm must cope with complex arithmetic. This
strategy is preferred to the direct use of complex numbers since the efficiency
of complex arithmetic is lower than real arithmetic[1]. However, it is possible to
consider the complex version of the algorithms by a straight transformation
in which the inner-product u^*v of two complex vectors u and v is defined by
$\bar{u}^T v$. In `MATLAB`, the operation is obtained by: `u' * v`.
Since the Hessenberg structure of a matrix plays an important role in the QR
method, it is important to see how any matrix A can be similarly reduced to
a Hessenberg matrix.

5.2.1 Reduction to an Upper Hessenberg Matrix

Let us first give the following definition.

Definition 5.3 *A square matrix $A \in \mathbb{R}^{n \times n}$ is upper Hessenberg, when*

$$A_{ij} = 0, \ for \ j = 1, \cdots, n-2, \quad i = j+2 \cdots, n.$$

Moreover, the matrix is unreduced if and only if $A_{i+1,i} \neq 0$, for $i = 1, \cdots, n-1$.

The Householder reflections are introduced in Section 4.4.1. They are symmetric matrices of the form

$$H = I - \beta vv^T \text{ where } \beta = 2/\|v\|^2. \tag{5.8}$$

The matrix H is never assembled since to compute Hx, this is performed by

$$Hx = x - \beta(x^T v) \ v.$$

In Section 4.4.2, it was shown how by applying on the left side of A successive
Householder transformations H_1, H_2, $\cdots$, it is possible to finally get a triangular matrix. Here the process is adapted to build a sequence of orthogonally
similar matrices

$$A_0 = A, \text{ and } A_{k+1} = H_k A_k H_k, \text{ for } k = 1, 2, \cdots, n-2, \tag{5.9}$$

---

[1]One complex addition involves two real additions but one complex multiplication involves four real multiplications and two real additions. Therefore on average, one complex
operation corresponds to four real operations.

where H_k is a Householder transformation for $k \geq 1$ and such that the last matrix A_{n-2} is upper Hessenberg. Let us illustrate the process on a small matrix of order $n = 6$. The first step must be such that the structure of $A_1 = H_1 A H_1$ is

$$A_1 = \begin{pmatrix} \star & \star & \star & \star & \star & \star \\ \star & \star & \star & \star & \star & \star \\ 0 & \star & \star & \star & \star & \star \\ 0 & \star & \star & \star & \star & \star \\ 0 & \star & \star & \star & \star & \star \\ 0 & \star & \star & \star & \star & \star \end{pmatrix} \tag{5.10}$$

For that purpose, we select w_1 such that $H_1 = I - \beta_1 w_1 w_1^T$ introduces zeros in rows 3 to n. This can be done by considering $w_1 = \begin{pmatrix} 0 \\ \tilde{w}_1 \end{pmatrix}$ where $\tilde{w}_1$ defines the Householder transformation $\tilde{H}_1 = I_{n-1} - \tilde{\beta}_1 \tilde{w}_1 \tilde{w}_1^T \in \mathbb{R}^{(n-1) \times (n-1)}$ which transforms the vector $(A_{21}, \cdots, A_{n1})^T$ into a vector $(\star, 0, \cdots, 0)^T$. Then we define

$$H_1 = \begin{pmatrix} 1 & 0 \\ 0 & \tilde{H}_1 \end{pmatrix}.$$

By construction, the structure of the matrix $U_1 A$ is already of the form of (5.10). By the special form of H_1, this structure is maintained when H_1 is applied on the right side to get the matrix A_1 which is similar to A.

For the second step, the same approach with $H_2 = \begin{pmatrix} I_2 & 0 \\ 0 & \tilde{H}_2 \end{pmatrix}$ where $\tilde{H}_2 = I_{n-2} - \tilde{\beta}_2 \tilde{w}_2 \tilde{w}_2^T \in \mathbb{R}^{(n-2) \times (n-2)}$ with w_2 defined such that the components 4 to n of the second column of A_1 are annihilated:

$$A_2 = \begin{pmatrix} \star & \star & \star & \star & \star & \star \\ \star & \star & \star & \star & \star & \star \\ 0 & \star & \star & \star & \star & \star \\ 0 & 0 & \star & \star & \star & \star \\ 0 & 0 & \star & \star & \star & \star \\ 0 & 0 & \star & \star & \star & \star \end{pmatrix} \tag{5.11}$$

The algorithm can be iteratively applied to the columns $k = 3, \cdots, n-2$ by reducing at each step by 1 the size of $\tilde{w}$. On exit A_{n-2} is an upper Hessenberg matrix such that $A = Q A_{n-2} Q^T$ where Q is the orthogonal matrix $Q = H_1 H_2 \cdots H_{n-2}$.

The complexity of this transformation is clearly $O(n^3)$, since the kth iteration involves $O((n-k)^2)$ operations. Depending on the problem being studied, the matrix Q is needed or not. When needed, the matrix can be either recursively assembled or implicitly kept by the list of vector $w_1, \cdots, w_{n-2}$. In the former case, the matrix is assembled from the right because it saves computation.

5.2.2 QR Algorithm for an Upper Hessenberg Matrix

When applied on an upper Hessenberg matrix H, the QR-algorithm in its basic version is mathematically expressed by the following sequences of matrices (H_k) and (Q_k):

$$
\begin{aligned}
H_1 &= H \text{ and } Q_1 = I \\
\text{For} \quad & k \geq 1, \\
& H_k = Q_{k+1} R_{k+1}, \text{ (QR factorization)} \\
& H_{k+1} = R_{k+1} Q_{k+1},
\end{aligned}
\tag{5.12}
$$

where the QR-factorization is defined in Chapter 4. It can be noticed that the matrices H_k are related by

$$
H_{k+1} = Q_{k+1}^T H_k Q_{k+1}. \tag{5.13}
$$

Therefore, this guarantees the following result.

Lemma 5.1 *The matrices H_k are orthogonally similar for $k \geq 1$ and are upper Hessenberg as well.*

Proof. To prove that the upper Hessenberg form is kept along the iterations (5.12), one may consider the QR-factorization by Householder transformations. In that situation, from the special profile of the Householder transformations, it is easy to see that the orthogonal factor Q_{k+1} is upper Hessenberg as well (see Exercise 4.5). This implies that the matrix $R_{k+1} Q_{k+1}$ is upper Hessenberg. ∎

It can be shown that the complexity of the QR-factorization of a Hessenberg matrix is $O(n^2)$ instead of $O(n^3)$ for a general matrix. But, the second step involves a matrix multiplication which gives a total $O(n^3)$ complexity at each iteration (5.12). A technique with a lower complexity can be implemented through Givens rotations (they are defined in Section 4.4.4). For this purpose, the following theorem is needed.

Theorem 5.9 (Implicit Q Theorem) *Let $A \in \mathbb{R}^{n \times n}$. Let us assume that $Q = [q_1, \cdots, q_n]$ and $V = [v_1, \cdots, v_n]$ are orthogonal matrices such that $H = Q^T A Q$ and $G = V^T A V$ are unreduced upper Hessenberg matrices. If $q_1 = v_1$, then $q_i = \pm v_i$ and $|H_{i,i-1}| = |G_{i,i-1}|$, for $i \geq 2$.*

Proof. See reference [33]. ∎

The consequence of this theorem is practically important. Let us consider the transformation $H_2 = Q_1^T H_1 Q_1$, where Q_1 is the orthogonal matrix such that the matrix $Q_1^T H_1$ is upper triangular. The QR factorization of H_1 can be performed by a sequence of Givens rotations $Rot_1, Rot_2, \cdots, Rot_{n-1}$, where, for $i = 1 : n-1$, $Rot(i, i+1, \theta_i)$ is an adequate rotation of angle θ_i applied on the left side on rows i and $i+1$. Therefore, from

$$
H_2 = Rot_{n-1}^T \cdots Rot_2^T (Rot_1^T H_1 Rot_1) Rot_2 \cdots Rot_{n-1}, \tag{5.14}
$$

we can guarantee that the matrices H_2 and $Rot_1^T H_1 Rot_1$ are orthogonally similar and their entries in position $(1,1)$ are equal. Their first columns only differ on the entries in position $(2,1)$ and $(3,1)$. Annihilating the entry $(3,1)$ in $Rot_1^T H_1 Rot_1$ can be done by a Givens rotation $\tilde{R}_2 = Rot(2,3,\phi_2)$. Applying this rotation on the two sides to obtain $\tilde{R}_2^T (Rot_1^T H_1 Rot_1) \tilde{R}_2$ creates a non-zero entry in $(4,2)$. It can then be annihilated by a Givens rotation $\tilde{R}_3 = Rot(3,4,\phi_3)$ and so on up to the end. It can be shown that the matrices H_2 and $\tilde{R}_{n-1}^T \cdots \tilde{R}_2^T (Rot_1^T H_1 Rot_1) \tilde{R}_2 \cdots \tilde{R}_{n-1}$ have the same first column and are upper Hessenberg. Therefore, from the Q-implicit theorem, the matrices $Rot_2 \cdots Rot_{n-1}$ and $\tilde{R}_2 \cdots \tilde{R}_{n-1}$ have the same columns possibly with the opposite sign. Since each transformation involves $O(n)$ operations the complexity of the whole process is $O(n^2)$.

Let us illustrate the situation for $n = 6$:

$$H_1 = \begin{pmatrix} * & * & * & * & * \\ * & * & * & * & * \\ 0 & * & * & * & * \\ 0 & 0 & * & * & * \\ 0 & 0 & 0 & * & * \end{pmatrix}$$

Step 1. $Rot_1 = Rot(1,2,\theta_1)$ such that the entry $(2,1)$ of $Rot_1^T H_1$ is annihilated. By applying Rot_1 on the right side, this creates a non-zero entry at position $(3,1)$:

$$H_1^{(2)} = Rot_1^T H_1 Rot_1 = \begin{pmatrix} * & * & * & * & * \\ * & * & * & * & * \\ + & * & * & * & * \\ 0 & 0 & * & * & * \\ 0 & 0 & 0 & * & * \end{pmatrix}$$

Step 2. Consider the rotation $\tilde{R}_2 = Rot(2,3,\phi_2)$ which annihilates the entry $(3,1)$ of the matrix $(\tilde{R}_2^T H_1^{(2)})$. In $H_1^{(3)} = \tilde{R}_2^T H_1^{(2)} \tilde{R}_2$ a new non-zero entry appears in position $(4,2)$:

$$H_1^{(3)} = \tilde{R}_2^T H_1^{(2)} \tilde{R}_2 = \begin{pmatrix} * & * & * & * & * \\ * & * & * & * & * \\ 0 & * & * & * & * \\ 0 & + & * & * & * \\ 0 & 0 & 0 & * & * \end{pmatrix}$$

Step 3. Consider the rotation $\tilde{R}_3 = Rot(3,4,\phi_3)$ which annihilates the entry $(4,2)$ of the matrix $(\tilde{R}_3^T H_1^{(3)})$. In $H_1^{(4)} = \tilde{R}_3^T H_1^{(3)} \tilde{R}_3$ a new non-zero entry appears in position $(5,3)$:

$$H_1^{(4)} = \tilde{R}_3^T H_1^{(3)} \tilde{R}_3 = \begin{pmatrix} * & * & * & * & * \\ * & * & * & * & * \\ 0 & * & * & * & * \\ 0 & 0 & * & * & * \\ 0 & 0 & + & * & * \end{pmatrix}$$

Step 4. Consider the rotation $\tilde{R}_4 = Rot(4, 5, \phi_4)$ which annihilates the entry (5,3) of the matrix $(\tilde{R}_4^T H_1^{(4)})$. The matrix $H_1^{(5)} = \tilde{R}_4^T H_1^{(4)} \tilde{R}_4$ is now upper Hessenberg:

$$H_1^{(5)} = \tilde{R}_4^T H_1^{(4)} \tilde{R}_4 = \begin{pmatrix} * & * & * & * & * \\ * & * & * & * & * \\ 0 & * & * & * & * \\ 0 & 0 & * & * & * \\ 0 & 0 & 0 & * & * \end{pmatrix}$$

As we claimed it previously, the matrix $H_1^{(5)}$ is essentially the same as the matrix H_2 in the QR iteration. This technique is often called *bulge chasing*. A similar approach is followed when the matrices are symmetric (see Section 5.3.1).

Remark 5.6 *In order to include the first transformation which transforms the matrix A into the upper Hessenberg matrix H_1, the QR-iteration (5.12) can be adapted into:*

$$\begin{aligned} &\textit{Compute } H_1 \textit{ and } Q_1 \textit{ such that } H_1 = Q_1^T A Q_1 \textit{ is upper Hessenberg,} \\ &\textit{For } k \geq 1, \quad H_{k+1} = Q_{k+1}^T H_k Q_{k+1}, \textit{ (QR-step by bulge chasing)} \end{aligned} \quad (5.15)$$

5.2.3 Convergence of the QR Method

The iterative transformation (5.12) builds a sequence of upper Hessenberg matrices $(H_k)_{k \geq 0}$ which are all orthogonally similar. It can be shown that, in the general situation, the first sub-diagonal of H_k becomes sparse when k goes to infinity. Mathematically, there may happen some special situations of no convergence when there are eigenvalues of equal modulus as for instance an orthogonal matrix for which the sequence is constant. Practically, this situation will not happen once the shifts will be introduced (see below the acceleration of convergence).

Convergence

In the general situation, upon convergence, the non-zero entries of the first subdiagonal are always isolated (i.e., they are separated by at least one zero in the subdiagonal). The accumulation points of the matrix sequence are quasi-upper triangular matrices where the diagonal blocks are either of order 1 or of order 2. They correspond to a real-Schur factorization of H_1 and therefore of the original matrix A. A block of order one corresponds to a real eigenvalue, whereas a block of order 2 corresponds to a pair of conjugate eigenvalues.

In order to express a convergence result, we consider the situation where the QR-iteration is applied in $\mathbb{C}$ with unitary transformations.

Theorem 5.10 (Convergence of the QR Algorithm) *Let $H \in \mathbb{C}^{n \times n}$ an unreduced upper Hessenberg matrix of which eigenvalues satisfy*

$$|\lambda_1| > |\lambda_2| > \cdots > |\lambda_n| > 0. \tag{5.16}$$

Under the assumption (5.16), when applied on H, the QR algorithm builds a sequence of upper-Hessenberg matrices that are unitarily similar to H and such that the sequence converges (modulo a unitary diagonal matrix) to an upper-triangular matrix T. The diagonal of T is $\mathrm{diag}(\lambda_1, \cdots, \lambda_n)$.

Proof. See reference [18]. A simpler proof for the real symmetric case is also given in reference [20]. ∎

Acceleration of the Convergence

Introduction of shifts in the process accelerates the process. At iteration k, a shift μ_k can be involved in the basic QR step (5.12) by:

$$
\begin{aligned}
& H_1 \text{ and } Q_1 \text{ are defined as in (5.15)} \\
& \text{For} \quad k \geq 1, \\
& \qquad H_k - \mu_k I = Q_{k+1} R_{k+1} \text{ (QR Factorization)}, \\
& \qquad H_{k+1} = R_{k+1} Q_{k+1} + \mu_k I,
\end{aligned}
\tag{5.17}
$$

and where the actual implementation is done through the bulge chasing technique as in (5.15). In the special situation where μ_k is an exact eigenvalue of H_k, it can be proved that the entry $(n, n-1)$ of H_{k+1} is zero and μ_k appears in the corner (n, n) of H_{k+1}. In such a situation, the matrix is deflated by one unit. This is a theoretical remark since the eigenvalue is just one of the unknowns of the problem. However, by selecting some approximation of an eigenvalue the convergence will be accelerated. The first idea consists in selecting $\mu_k = (H_k)_{nn}$. This works well when there is a real eigenvalue in the neighborhood of μ_k but it is necessary to cope with the possibility of conjugate pairs of eigenvalues. The second idea is to deflate consecutively with two conjugate shifts and to organize the computation to stay in real arithmetic. This is the explicit *double shift strategy*. For that purpose the last 2×2 block of H_k is considered. When the two eigenvalues of this block μ_1 and μ_2 are real, the technique of a single shift can be successively used with the two eigenvalues as successive shifts:

$$
\begin{aligned}
& H_k - \mu_k I = Q_{k+1} R_{k+1}, \text{ (QR factorization)}, \\
& H_{k+1} = R_{k+1} Q_{k+1} + \mu_1 I \\
& H_{k+1} - \mu_2 I = Q_{k+2} R_{k+2}, \text{ (QR factorization)} \\
& H_{k+2} = R_{k+2} Q_{k+2} + \mu_2 I.
\end{aligned}
\tag{5.18}
$$

In this case, it can be proved that (proof is left as Exercise 5.4):

$$(H_k - \mu_2 I)(H_k - \mu_1 I) = Q_{k+1} Q_{k+2} R_{k+2} R_{k+1}. \tag{5.19}$$

Therefore H_{k+2} can be directly obtained from H_k by considering the QR factorization of the matrix $G = (H_k - \mu_2 I)(H_k - \mu_1 I) = (H_k - \mu_1 I)(H_k - \mu_2 I)$. When the two eigenvalues of the block form a conjugate pair, the computation can be rearranged to maintain the real arithmetic:

$$(H_k - \mu_2 I)(H_k - \mu_1 I) = H_k^2 - \tau H_k + \delta I, \qquad (5.20)$$

where $\tau = 2\mathcal{R}e(\mu_1)$ and $\delta = |\mu_1|^2$ are respectively the trace and the determinant of the 2×2 block.

The drawback of this method comes from the need of evaluating the matrix G which involves $O(n^3)$ arithmetic operations. But here again, an implicit technique will be applied, the so-called *implicit double shift strategy*. However, it has to be tuned to the situation: the first column of G has now three non-zero entries. Therefore, instead of Givens rotations, the technique uses Householder transformations. The technique of bulge chasing based on Theorem 5.9 remains, and it leads to a complexity in $O(n^2)$ instead of $O(n^3)$ for the arithmetic operations. This is the Francis QR step. For a precise description, see, for example, reference [33].

5.3 Algorithms for Symmetric Matrices

In this section, the QRs method is adapted to the special situation when the matrix A is symmetric.

5.3.1 Reduction to a Tridiagonal Matrix

If we apply the orthogonal transformations which transform the matrix A into an upper Hessenberg matrix as indicated in (5.9) when A is symmetric, it is easy to see that all the matrices A_k for $k = 1, \cdots, n - 2$ are symmetric as well. Therefore the last matrix $T = A_{n-2}$ is a symmetric upper Hessenberg matrix. Such a matrix is a symmetric tridiagonal matrix.

When applying the successive orthogonal transformations U_k, it is possible to take into account the symmetry of the matrices A_k to lower the number of arithmetic operations. This can be done by mean of the procedure of BLAS2 that updates a matrix with a rank 2 update: $M \to M + VV^T$ where $V \in \mathbb{R}^{n \times 2}$. Such an approach is implemented in the procedure DSYTRD of the LAPACK library [5].

When the matrix $Q = H_1 \cdots H_{n-2}$ is required, the procedure is as in the general case.

5.3.2   Algorithms for Tridiagonal Symmetric Matrices

When the matrix is symmetric, the eigenvalues are real and the eigenvectors form an orthonormal basis of $\mathbb{R}^n$. These properties are used to find robust algorithms to compute the eigenvalues of a symmetric tridiagonal matrix involving less arithmetic operations than in the general case for Hessenberg matrices. The first approach is based on the interlacing property as stated in Theorem 5.7.

Unreduced Matrices and Sturm Sequences

Let $T \in \mathbb{R}^{n \times n}$ be a symmetric tridiagonal matrix. We denote $T = [\beta_i, \alpha_i, \beta_{i+1}]_{1,n}$ where $T_{i,i} = \alpha_i$ for $i = 1, \cdots, n$ and $T_{i,i-1} = \beta_i$ for $i = 2, \cdots, n$. For $k = 1, \cdots, n$, the characteristic polynomial of the principal matrix T_k of T is $p_k(\lambda) = \det(T_k - \lambda I_k)$.
Let us assume that T is unreduced, i.e., $\prod_{k=2}^{n} \beta_k \neq 0$ or otherwise, T would be structured in at least two diagonal blocks, each of them being tridiagonal symmetric and the eigenvalue problem would be a collection of smaller tridiagonal eigenvalue problems.

Proposition 5.5 *The sequence of the characteristic polynomials $p_k(\lambda)$ of the principal submatrices T_k of T satisfies the recursion*

$$\begin{cases} p_0(\lambda) = 1, \ p_1(\lambda) = \alpha_1 - \lambda \ and \ for \ k \geq 1, \\ p_{k+1}(\lambda) = (\alpha_{k+1} - \lambda)p_k(\lambda) - \beta_{k+1}{}^2 p_{k-1}(\lambda). \end{cases} \tag{5.21}$$

This sequence is the Sturm sequence in λ.

Proof. The proposition is proved by induction. It is obviously true for $k = 1$. Let us assume that the property is satisfied up to k. At the order $k + 1$, the polynomial $p_{k+1}(\lambda) = \det(T_{k+1} - \lambda I_{k+1})$ can be decomposed by developing the last column of $T_{k+1} - \lambda I_{k+1}$. ∎

Corollary 5.3 *The eigenvalues of an unreduced symmetric tridiagonal matrix are simple eigenvalues.*

Proof. For any $\lambda \in \mathbb{R}$, the sequence $p_k(\lambda)$ for $k = 1, \cdots, n$ cannot include two null consecutive terms. If there were two such terms $p_{k+1}(\lambda) = p_k(\lambda) = 0$ for some k, and because $\beta_i \neq 0$ for $i = 1, \cdots, k$, that would imply $p_0(\lambda) = 0$ which is not true.
The corollary can then be proved by induction. At order 2, it is easy to see that if λ is a double eigenvalue of T_2 then $T_2 = I_2$ which is not an unreduced matrix. Let us assume that any unreduced symmetric tridiagonal matrix of order $k \geq 2$ has only simple eigenvalues. Let T_{k+1} be unreduced symmetric tridiagonal of order $k + 1$. Its principal submatrix T_k of order k has k simple eigenvalues by the assumption of the induction. From the interlacing property (Theorem 5.7), these eigenvalues separate the eigenvalues of T_{k+1}. The separation is

strict since, as mentioned at the beginning of the proof, p_k and p_{k+1} have no common roots. This ends the proof.   ∎

The following theorem provides an effective technique to localize the eigenvalues of a symmetric tridiagonal matrix.

Theorem 5.11 (Sturm Sequence Property) *Let $T \in \mathbb{R}^{n \times n}$ be an unreduced symmetric tridiagonal matrix. With the previous notations, and for $k = 1, \cdots, n$ and $\lambda \in \mathbb{R}$, the sign of $p_k(\lambda)$ is defined by*

$$\text{sgn } (p_k(\lambda)) = \begin{cases} \text{sign of } p_k(\lambda) \text{ when } p_k(\lambda) \neq 0, \\ \text{sign of } p_{k-1}(\lambda) \text{ when } p_k(\lambda) = 0. \end{cases} \tag{5.22}$$

Let $N(\lambda)$ be the number of sign changes between consecutive terms of the sequence $(\text{sgn } p_k(\lambda))_{(k=0,\cdots,n)}$.
$N(\lambda)$ is equal to the number of eigenvalues of T that are smaller than λ.

Proof. See, for example, reference [20].   ∎

This result provides an effective technique to localize the eigenvalues lying in a given interval $[a, b]$: the number $N(b) - N(a)$ provides the number of eigenvalues that must be computed; then by successive bisections of the interval, all the eigenvalues can be localized in intervals at the desired accuracy. This technique involves $O(n)$ arithmetic operations for each bisection.

In order to avoid appearance of overflows or underflows, instead of considering the basic Sturm sequence, a transformed one is computed: the sequence of the consecutive ratios $q_k(\lambda) = \frac{p_k(\lambda)}{p_{k-1}(\lambda)}$ can be given from $k = 1$ to $k = n$ by the following recursion:

$$\begin{cases} q_1(\lambda) = \alpha_1 - \lambda, \text{ and for } k \geq 1, \\ q_{k+1}(\lambda) = \alpha_{k+1} - \lambda - \frac{\beta_{k+1}^2}{q_k(\lambda)}. \end{cases} \tag{5.23}$$

The number $N(\lambda)$ is equal to the number of negative terms of the sequence $(q_k(\lambda))_{k=1,\cdots,n}$. It can be shown that this sequence is actually the diagonal of the matrix U of the LU-factorization of the matrix $T - \lambda I$ (see Exercise 5.5). Mathematically, the function $N(\lambda)$ is a nondecreasing function; however, roundoff errors may perturb this property as indicated in reference [23]. This situation is very uncommon.

Once the eigenvalues are computed, the corresponding eigenvectors can be obtained by Inverse Iterations (Algorithm 5.2). For clusters of eigenvalues (i.e., very close eigenvalues), the obtained eigenvectors must be mutually reorthogonalized. This is done by applying the Modified Gram-Schmidt orthogonalization (Algorithm 4.3). A more sophisticated technique was introduced in reference [37].

QR Algorithm

The QR method, which has been introduced in Section 5.2.2 for Hessenberg matrices, can be adapted to the special situation of a symmetric tridiagonal matrix. The two following propositions are obtained by a straightforward application of the general case.

Proposition 5.6 *Let $T \in \mathbb{R}^{n \times n}$ be a symmetric tridiagonal matrix, and $\lambda \in \mathbb{R}$. By QR factorization, there exist an upper Hessenberg orthogonal matrix Q and an upper triangular matrix R such that: $T - \mu I = QR$.*
Then, the matrix $T_1 = RQ + \mu I$ is a symmetric tridiagonal matrix orthogonally similar to T.

Proof. Since Q and RQ are upper Hessenberg matrices, the matrix $T_1 = RQ + \mu I = Q^T T Q$ is upper Hessenberg and symmetric. Therefore, it is tridiagonal. ∎

Proposition 5.7 *With the previous notations, if T is an unreduced matrix and if μ is an eigenvalue of T, then the last column of T_1 is the vector μe_n (where e_n is the last vector of the canonical basis).*

The QR method can be applied in the same way as for general Hessenberg matrices. For a symmetric matrix, there is no need to consider a double shift since all the eigenvalues are real. The explicit shift version writes:

$$
\begin{aligned}
&T_0 = T \text{ and } Q_0 = I_n \\
&\text{For} \quad k \geq 0, \\
&\qquad \text{Define the shift } \mu_k, \\
&\qquad T_k - \mu_k I = Q_{k+1} R_{k+1} \text{ (QR factorization)}, \\
&\qquad T_{k+1} = R_{k+1} Q_{k+1} + \mu_k I.
\end{aligned}
\tag{5.24}
$$

An *implicit shift* version is often preferred (for example, see [33]). The complexity of the procedure is low: $O(n)$ operations per iteration when the matrix of eigenvectors is not required; when the matrix Q_k is assembled, the complexity of one iteration is $O(n^2)$.

At each iteration, the shift μ_k must be selected. The situation is simpler than for Hessenberg matrices since there is only one shift to define. The first approach consists in selecting the last diagonal entry of T_k: $\mu_k = (T_k)_{nn}$. Often, a better shift is chosen by selecting the eigenvalue of the last 2×2 block that is closest to $(T_k)_{nn}$. It is called the *Wilkinson shift*.

The convergence of the method is expressed by the fact that the matrix T_k becomes diagonal when k goes to infinity; i.e., $\lim_{k \to \infty} \text{Off}(T_k) = 0$. It is proved that the two shifts $\mu_k = (T_k)_{nn}$ and the Wilkinson shift imply a cubic convergence of the method.

5.4   Methods for Large Size Matrices

When the matrix is very large and sparse, it is very difficult and often impossible to transform it into a dense Hessenberg matrix. Therefore, in this section, we consider methods that only use the matrix A as an operator: the only available operation is the matrix-vector multiplication $v \longrightarrow Av$.

5.4.1 Rayleigh-Ritz Projection

Let us assume that the columns of $V = [v_1, \cdots, v_k] \in \mathbb{R}^{n \times k}$ form an orthonormal basis of a k-dimensional invariant subspace $\mathcal{X}$ of the matrix $A \in \mathbb{R}^{n \times n}$. Therefore

$$V^*V = I_k, \text{ and}$$
$$\forall i = 1, \cdots, k, \ \exists y_i \in \mathbb{R}^k, \text{ such that } Av_i = Vy_i.$$

An orthonormal basis U of the entire space $\mathbb{R}^{n \times n}$ is obtained by adjoining a set of vectors $W \in \mathbb{R}^{n \times (n-k)}$ to V to get the orthogonal matrix $U = [V, W]$. By expressing the operator defined by A into the basis U, the new matrix is:

$$U^T A U = \begin{pmatrix} V^T A V & V^T A W \\ O & W^T A W \end{pmatrix}. \tag{5.25}$$

We notice that the block $W^T A V$ is null since the columns of AV belong to the invariant subspace $\mathcal{X}$ and therefore are orthogonal to $\mathcal{X}^{\perp}$. If the matrix A is Hermitian, the block $V^T A W$ is null as well.

The block-triangular form of (5.25) shows that the set of eigenvalues of A is equal to the union of the set of the k eigenvalues of $H = V^T A V$ and the set of the $n - k$ eigenvalues of $W^T A W$. The matrix $H \in \mathbb{R}^{k \times k}$ can be seen as the matrix of the restriction of the operator to the subspace $\mathcal{X}$ when expressed in the basis V.

Let us now consider that the subspace $\mathcal{X}$ is not invariant for A. Therefore, the block $W^T A V$ is not zero. By decomposing the matrix AV on the direct sum $\mathbb{R}^{n \times n} = \mathcal{X} \oplus \mathcal{X}^{\perp}$, we get:

$$AV = VV^T(AV) + (I - VV^T)(AV) \tag{5.26}$$
$$= VH + R, \tag{5.27}$$

where $R = AV - VH = (I - VV^T)(AV)$. If the subspace $\mathcal{X}$ is near an invariant subspace, the quantity $\|R\|_2$ is almost zero. This quantity is a measure of the quality of the approximation of the invariance. For the special case of a Hermitian matrix A, the following theorem provides enclosures for the eigenvalues.

Theorem 5.12 *Let $A \in \mathbb{R}^{n \times n}$ be a symmetric matrix, and $(\lambda_i)_{i=1,\cdots,n}$ its eigenvalues. Let $V \in \mathbb{R}^{n \times k}$ such that $V^*V = I_k$. If $(\mu_i)_{i=1,\cdots,k}$ denote the eigenvalues of the matrix $H = V^*AV$, then there exist k eigenvalues $(\lambda_{i_j})_{j=1,\cdots,k}$ of A such that*

$$|\lambda_{i_j} - \mu_j| \leq \|R\|_2$$

where $R = AV - VH$.

Proof. Let us first remark that

$$R^T V \; = \; 0. \tag{5.28}$$

Therefore

$$\left(A - (RV^T + VR^T) \right) V \;=\; VH. \tag{5.29}$$

It can be proved (see Exercise 5.6) that

$$\| RV^T + VR^T \|_2 \;=\; \| R \|_2. \tag{5.30}$$

The final result is an application of Theorem 5.6. ∎

More precisely, when the orthonormal basis V of an approximate invariant subspace is considered, the full eigendecomposition of the matrix $H = V^T A V$ can be computed. For every eigenpair (μ, y) of H, an approximated eigenpair of A is obtained by the corresponding Ritz pair:

Definition 5.4 *With the previous notations, the pair (μ, x) where $x = Vy$ is called Ritz pair of A. The Ritz value μ is equal to the Rayleigh quotient corresponding to the Ritz vector x:*

$$\mu = \rho_A(x) = x^T A x = y^T H y.$$

The computation of the residual of a Ritz pair (μ, x) is easy to evaluate:

$$\begin{aligned} r &= Ax - \mu x \\ &= AVy - \mu Vy \\ &= V(Hy - \mu y) + Ry \\ &= Ry \end{aligned}$$

and the norm of the residual is $\|r\|_2 = \|Ry\|_2 \le \|R\|_2$.

5.4.2 Arnoldi Procedure

Many algorithms for solving problems that involve a large sparse matrix A are based on Krylov subspaces.

Definition 5.5 *Given the matrix $A \in \mathbb{R}^{n \times n}$ and a non-zero initial vector $v \in \mathbb{R}^n$, the $k-th$ Krylov subspace is the subspace of $\mathbb{R}^n$ spanned by the $k-1$ successive powers of A applied to v:*

$$\mathcal{K}_k(A, v) = span(v, Av, A^2 v, \cdots, A^{k-1} v).$$

Therefore this set can be seen as the set $\mathcal{P}_{k-1}(A)(v)$ of the polynomials of degree smaller than k applied to the vector v.

Clearly, the dimension of $\mathcal{K}_k(A, v)$ cannot be greater than k. When $\dim(\mathcal{K}_k(A, v)) < k$, it can be shown that the Krylov subspace is an invariant subspace for A (see Exercise 5.8). Let us now consider the situation where $\dim(\mathcal{K}_k(A, v)) = k$.

In order to build an orthonormal basis of the Krylov subspace, we proceed by induction:

- First step: compute $v_1 = \frac{v}{\|v\|}$;

- Induction: let us assume that, for $k \geq 1$, the vectors $v_1, \cdots, v_k$ are already known and are such that $v_k \notin \mathcal{K}_{k-1}(A, v)$. The system of vectors $V_k = [v_1, v_1, v_2, \cdots, v_k]$ is an orthonormal basis of $\mathcal{K}_{k-1}(A, v)$. Therefore, the system of vectors $[v_1, v_1, v_2, \cdots, v_{k-1}, Av_k] = [V_k, Av_k]$ spans $\mathcal{K}_{k+1}(A, v)$. This is a basis of this space, otherwise the dimension of the Krylov subspace $\mathcal{K}_{k+1}(A, v)$ would be smaller than $k + 1$. The vector v_{k+1} is then obtained by successively projecting it onto the orthogonal of the vectors $v_1, \cdots, v_k$ and normalizing then the result:

$$
\begin{aligned}
w &= (I - v_k v_k^T) \cdots (I - v_1 v_1^T) A v_k, \\
v_{k+1} &= \frac{w}{\|w\|}.
\end{aligned}
$$

The resulting procedure is given in Algorithm 5.4.

Algorithm 5.4 Arnoldi Procedure

```
function [V,H] = arnoldi(A,m,v)
% Input:
%    A: matrix
%    m: dimension of the Krylov subspace
%    v: initial vector
% Output V(:,1:m+1) and H(1:m+1,1:m) such that: A*V(:,1:m) = V*H
n=size(A,1) ; V=zeros(n,m+1) ; H=zeros(m+1,m) ;V(:,1)=v/norm(v);
for k=1:m,
    w = A*V(:,k) ;
    for j=1:k,
       H(j,k) = V(:,j)'*w ;w = w - H(j,k)*V(:,j) ;
    end ;
    H(k+1,k) = norm(w);V(:,k+1) = w/H(k+1,k) ;
end
```

This algorithm is based on BLAS-1 operations and it involves multiplications of vectors by the matrix A.

It can be seen that the Arnoldi procedure corresponds to apply the

Modified Gram-Schmidt process (see Algorithm 4.3) on the matrix $[v_1, Av_1, Av_2, \cdots, Av_k] = [v_1, AV_k]$.

Theorem 5.13 (Arnoldi Relations) *Algorithm 5.4 builds the orthogonal basis $V_{k+1} \in \mathbb{R}^{n \times (k+1)}$ of the Krylov subspace $\mathcal{K}_{k+1}(A, v)$ and the upper-Hessenberg matrix $\tilde{H}_k = V_{k+1}^T AV_k \in \mathbb{R}^{(k+1) \times k}$ that satisfy the following relations:*

$$AV_k = V_{k+1}\tilde{H}_k, \tag{5.31}$$
$$AV_k = V_k H_k + \beta_{k+1} v_{k+1} e_k^T, \tag{5.32}$$
$$V_k^T AV_k = H_k, \tag{5.33}$$

where the matrix $\tilde{H}_k$ is partitioned into

$$\tilde{H}_k = \left(\begin{array}{c|c} & H_k \\ \hline 0 \quad \cdots \quad 0 & \beta_{k+1} \end{array} \right), \tag{5.34}$$

and where e_k is the $k-th$ canonical vector of $\mathbb{R}^k$.

Proof. Since the Arnoldi process corresponds to apply the MGS algorithm to the columns of the matrix $[v, Av_1, \cdots, Av_k]$, it performs a QR factorization of this matrix:

$$[v_1, Av_1, \cdots, Av_k] = [v_1, v_2, \cdots, v_{k+1}]R_{k+1}.$$

The first column of R_{k+1} is $\|v\|e_1^{(k+1)}$ where $e_1^{(k+1)}$ is the first canonical vector of $\mathbb{R}^{k+1}$. Let us define $\tilde{H}_k$ by the remaining columns of R_{k+1}. Therefore, it becomes obvious that $\tilde{H}_k \in \mathbb{R}^{(k+1) \times k}$ is an upper Hessenberg matrix and that (5.31) holds. Relations (5.32) and (5.33) are direct consequences of the partition of $\tilde{H}_k$ as expressed in (5.34). ∎

The number of floating point operations involved in the Arnoldi procedure for building V_{m+1} is $2mn_z + 2m^2 n + O(mn)$, where n_z is the number of non-zero entries of A. Clearly, the maximum dimension m of the Krylov subspace is limited (i) by the storage volume needed by V_{m+1}, and (ii) by the number of operations which increases with $m^2 n$.

5.4.3 The Arnoldi Method for Computing Eigenvalues of a Large Matrix

The formula (5.33) illustrates that H_k can be interpreted as being a projection of the operator A when restricted to the Krylov subspace $\mathcal{K}_k(A, v)$. Comparing the formula (5.32) to the formula (5.27) with $V = V_k$, we get $R = \beta_{k+1} v_{k+1} e_k^T$ which is the defect of invariance of the subspace. Since $R^T R = \beta_{k+1}^2 e_k e_k^T$, we obtain a measure of the lack of invariance with $\|R\|_2 = \beta_{k+1}$.

The Ritz values obtained with the subspace $\mathcal{K}_k(A, v)$ (i.e., the eigenvalues of H_k) can be considered as approximations of some eigenvalues of A.

Proposition 5.8 *With the previous notations, let (μ, y) be an eigenpair of H_k. The corresponding Ritz pair is (μ, x) where $x = V_k y$. The corresponding residual is:*

$$r = Ax - \mu x = \beta_{k+1} \nu_k v_{k+1}, \tag{5.35}$$

where ν_k is the last component of y; therefore,

$$\|r\|_2 = \beta_{k+1} |\nu_k|. \tag{5.36}$$

Proof. Left as Exercise 5.9. ∎

In order to illustrate how the Ritz values approximate eigenvalues of A when the dimension of the Krylov subspace increases, we plot in Figure 5.1, the situation with the matrix `BFW398A`, a matrix of order $n = 398$, obtained from the Matrix Market set of test matrices [11]. For that purpose, two dimensions of the Krylov subspace are considered ($k = 10$ and $k = 70$) with a random initial vector. The Ritz values approximate first the most peripheral eigenvalues in the spectrum. The non-real eigenvalues are not approximated when the dimension of the Krylov subspace is too small. In Figure 5.1, the Ritz values are obtained from Krylov subspaces (Arnoldi procedure on the matrix `BFW398A` for $k = 10$ and $k = 70$; the eigenvalues are marked with dots and the Ritz values are marked with circles; the scales are non-equal in x and y).

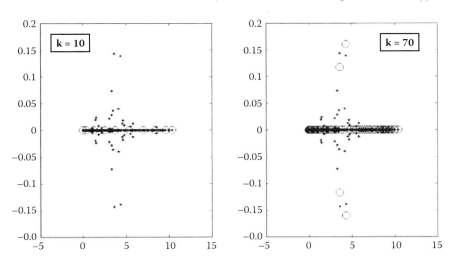

FIGURE 5.1: Approximating eigenvalues by Ritz values

The convergence of the Ritz values to the eigenvalues was studied but it is too difficult for the scope of this book. For a presentation, the reader may consider [57].

5.4.4   Arnoldi Method for Computing an Eigenpair

Let us assume that we search an eigenpair (λ, x) of A where $|\lambda|$ is maximum. The dimension k of the Krylov basis must be large enough to get valid approximations of the eigenvalues of A but two obstacles imply limitations on k:

- the storage V_{k+1} is limited by the memory capacity. This matrix is dense and therefore implies storage of $n(k+1)$ words. When n is huge, k must be smaller than some value m.

- roundoff errors generate a loss of orthogonality. When k is too large the computed system $fl(V_k)$ is not orthogonal and often is even not a full column rank matrix.

The Iterative Method with Explicit Restart

Let us assume that the maximum dimension of the Krylov subspace is m. Except for a well chosen initial vector, it is likely the situation where the corresponding Ritz pair does not approximate an eigenvalue of A with a sufficient accuracy. A restarting strategy is needed: a new m-dimensional Krylov subspace will be built but with a better initial vector. The simplest method consists of choosing as starting vector the best approximation to the sought eigenvector. This technique is expressed in Algorithm 5.5. In this algorithm, the residual is computed from its definition and not by using the relation (5.36) which would have saved one multiplication by A at every iteration. This is to cope with a possible loss of orthogonality which would invalidate relation (5.36).

Algorithm 5.5 Arnoldi Method for Computing the Dominant Eigenpair

```
function [x,lambda]=arnoldeigs(A,v,m,tol)
%  [x,lambda]=arnoldeigs(A,v,m,tol)
%  Arnoldi for one eigenvalue of largest modulus
itmax=1000 ; n=max(size(x));
x=v/norm(v) ; tolA= tol*norm(A,inf) ;
converge=0 ; iter=0 ; nmult=0 ;
while ~converge  & iter < itmax,
    iter=iter+1 ;
    [V,H] = arnoldi(A,x,m) ;
    nmult=nmult+m ;
    [Z,D]=eig(H(1:m,1:m)); D=diag(D) ;
    [lamax,k]=max(abs(D)) ; lambda=D(k) ; z=Z(:,k);
    x=V(:,1:m)*z ; r=A*x-lambda*x ;
    converge = norm(r) < tolA ;
end ;
```

If the computed eigenvalue λ is not real, then $\bar{\lambda}$ is also an eigenvalue since A is a real matrix. The corresponding eigenvector is $\bar{x}$ where x is the eigenvector of λ. It is possible to define a real basis of a 2-dimensional invariant subspace such that the corresponding Ritz values are λ and $\bar{\lambda}$ (see Exercise 5.7).

Deflation

When several eigenvalues are sought, a *deflation* is considered. Let us assume that one real eigenpair (λ, x) is already known. To compute another eigenvalue of large modulus, the Arnoldi method is now applied on the matrix $B = (I - xx^T)A$ instead of A, since the spectrum of B is $\Lambda(B) = (\Lambda(A)\backslash\{\lambda\})\cup\{0\}$ (see Exercise 5.11). Let us assume that (μ, y) is an eigenpair of B: $(I - xx^T)Ay = \mu y$. Hence the system $X = [x, y]$ is an orthonormal basis of a 2-dimensional invariant subspace:

$$AX = X \begin{pmatrix} \lambda & \nu \\ 0 & \mu \end{pmatrix}, \tag{5.37}$$

where $\nu = x^T A y$. Equation (5.37) can be interpreted as a partial Schur factorization (i.e., limited to an invariant subspace). The eigenvector z corresponding to μ is obtained from the Ritz vector $z = Xt$ where t is the eigenvector of the matrix $\begin{pmatrix} \lambda & \nu \\ 0 & \mu \end{pmatrix}$ which corresponds to μ. The same technique can be repeated by searching the eigenvalue of largest modulus of $(I - XX^T)A$ and so on. Step by step, more eigenpairs can be found.

The `ARPACK` Method

The `ARPACK` software [45] implements the Implicitly Restarted Arnoldi Method (IRAM). It corresponds to the iterative Arnoldi method as described in this chapter except for its restarting strategy. This method is the one implemented in the MATLAB function `eigs`.

The implicit restarting technique is a technique which compresses the Krylov subspace of maximum allowed dimension m to a small dimension k where k represents the number of sought eigenvalues. Let us assume that the m Ritz values obtained from the last Hessenberg matrix H_m are partitioned into k wanted and $m - k$ unwanted values. The technique is based on $m - k$ steps of the QR method applied on H_m using the unwanted Ritz values as shifts. For a complete description, see references [26, 46].

5.4.5   Symmetric Case: Lanczos Algorithm

When $A \in \mathbb{R}^{n \times n}$ is symmetric, the Hessenberg matrix appearing in (5.32-5.33) is symmetric as well. Since the matrix is an upper Hessenberg matrix, the matrix is symmetric and tridiagonal. Instead of H_k and $\tilde{H}_k$, we denote these matrices, respectively, by T_k and $\tilde{T}_k$ to remember their special pattern:

$$\tilde{T}_k = \begin{pmatrix} \alpha_1 & \beta_2 & & & & & \\ \beta_2 & \alpha_2 & \beta_3 & & & & \\ & \beta_3 & \alpha_3 & \beta_4 & & & \\ & & . & . & . & & \\ & & & . & . & \beta_k \\ & & & & \beta_k & \alpha_k \\ & & & & & \beta_{k+1} \end{pmatrix}, \tag{5.38}$$

and the relations (5.31-5.33) become

$$AV_k = V_{k+1}\tilde{T}_k \tag{5.39}$$

$$V_k^T AV_k = = T_k \tag{5.40}$$

$$AV_k = V_k T_k + \beta_{k+1} v_{k+1} e_k^T, \tag{5.41}$$

From (5.39), an important consequence is that the three-term recurrence is satisfied by the columns of the matrix V_k:

$$Av_k = \beta_k v_{k-1} + \alpha_k v_k + \beta_{k+1} v_{k+1}. \tag{5.42}$$

This implies that mathematically, the modified Gram-Schmidt procedure which is involved in the Arnoldi procedure is simplified: Algorithm 5.4 becomes Algorithm 5.6.

Algorithm 5.6 Lanczos Procedure

```
function [T,V] = lanczos(A,v,m)
%    [T,V]=lanczos(A,v,m)  or  [T]=lanczos(A,v,m)
%    Tridiagonalization without reorthogonalization
%    Basis is stored only if V is requested ;
%    T is a (m+1)xm symmetric tridiagonal matrix
%    V (optional): Krylov basis
 store =   nargout == 2 ;
  n=size(v,1) ; T=sparse(m+1,m) ; v=v/norm(v) ;
  if store,   V=[v] ; end ;
  for k=1:m,
        w = A*v ;
    if k>1,
      T(k-1,k)=T(k,k-1);w=w-vold*T(k,k-1);
    end ;
    T(k,k)=v'*w ;w=w-v*T(k,k) ;
    T(k+1,k)=norm(w) ;vold=v; v=w/T(k+1,k);
    if store, V=[V,v] ; end ;
  end ;
end
```

The Lanczos procedure has a lower complexity than the Arnoldi procedure: the number of floating point operations involved is $2mn_z + 9mn + O(n)$ and not $2mn_z + 2m^2n + O(mn)$ as in the Arnoldi procedure. Combined with the fact that the procedure can be run without storing the whole Krylov basis V_m, it becomes possible to run the algorithm for very large m.

Computing Eigenvalues with Lanczos

As just mentioned, by applying Lanczos on a symmetric matrix A of order n, one should be able to build the corresponding tridiagonal matrix T_m for a very large m as for instance $m = n$. This would bypass the need of restarting techniques as it is done with Arnoldi.

Practically, a loss of orthogonality appears quickly when one eigenvector of A becomes close to belonging to the Krylov subspace. This phenomenon has been studied by Paige [50, 51, 52]. When it occurs, not only orthogonality is lost but even the linear independency of the columns of V_m is not insured. In that situation, there are duplications of the Ritz values.

To remedy the loss of orthogonality, one may consider:

- *Lanczos with complete reorthogonalization:* every v_{k+1} is reorthogonal-ized against all the previous v_i, for $i = 1, \cdots, k$. This is the Arnoldi procedure. It implies to store the matrix V_m and to consider a restart-ing startegy.

- *Lanczos with partial reorthogonalization:* every v_{k+1} is reorthogonalized with all the previous v_i, for $i = 1, \cdots, k$ but only when necessary. This is a way to skip some computations in Arnoldi but restarting is still necessary.

- *Selective reorthogonalization:* every v_{k+1} is reorthogonalized against all the eigenvectors already computed. This technique involves less arith-metic operations than the previous ones. The method is described in reference [53].

- *No orthogonalization:* if no re-orthorgonalizing technique is included, the Ritz values appear under several copies for only one corresponding eigenvalue of A. It is therefore necessary to include a technique to discard the unwanted copies. This approach is considered in reference [21]. See also Exercise 5.12.

The `ARPACK` method designed for the non-symmetric case can also be used with symmetric matrices: this is the Lanczos with full re-orthogonalization with implicit restart. The code is simplified since all the eigenvalues are real. This is the method used by the function `eigs` of `MATLAB`.

5.5   Singular Value Decomposition

This decomposition is defined by Definition 2.14 and Theorem 2.7. Therefore, it can be obtained from the diagonalization of $A^T A$ or of AA^T by choosing the smallest dimension when A is a rectangular matrix. An alternative approach can also be considered with the following theorem.

Theorem 5.1 *Let $A \in \mathbb{R}^{m \times n}$ $(m \geq n)$ have the singular value decomposition*

$$U^\top AV = S,$$

where $U = [u_1, \cdots, u_m] \in \mathbb{R}^{m \times m}$ and $V = [v_1, \cdots, v_n] \in \mathbb{R}^{n \times n}$ are orthogonal matrices and $S = \begin{pmatrix} \Sigma \\ 0 \end{pmatrix} \in \mathbb{R}^{m \times n}$ is a rectangular matrix where $\Sigma = \mathrm{diag}(\sigma_1, \cdots, \sigma_n) \in \mathbb{R}^{n \times n}$. The symmetric matrix

$$A_{aug} = \begin{pmatrix} 0_m & A \\ A^\top & 0_n \end{pmatrix} \in \mathbb{R}^{(m+n) \times (m+n)} \tag{5.43}$$

is called the augmented matrix. For $i = 1, \cdots, n$, the scalars σ_i and $-\sigma_i$ are eigenvalues of A_{aug} corresponding respectively to the eigenvectors $\frac{\sqrt{2}}{2} \begin{pmatrix} u_i \\ v_i \end{pmatrix}$ and $\frac{\sqrt{2}}{2} \begin{pmatrix} u_i \\ -v_i \end{pmatrix}$. There are $m-n$ additional zero eigenvalues to get the whole spectrum of A.

Proof. Left as an exercise (Exercise 5.13). ∎

Therefore, this augmented matrix A_{aug} provides a second way to compute the singular value decomposition in addition to the normal matrix $A_{nrm} = A^T A$.

5.5.1   Full SVD

Let us consider the situation where all the singular values of the matrix $A \in \mathbb{R}^{m \times n}$, with $m \geq n$, are sought. The way to proceed is then to implicitly apply to $A_{nrm} = A^T A$ the approach that was developed for computing the eigenvalues of a symmetric matrix.

Bidiagonalization

In order to implicitly tridiagonalize the matrix A_{nrm}, a technique will consist in determining two orthogonal matrices $U_0 \in \mathbb{R}^{m \times m}$ and $V_0 \in \mathbb{R}^{n \times n}$ such that the matrix $B = U_0^T A V_0$ is upper bidiagonal

$$
B \;=\;
\begin{pmatrix}
\alpha_1 & \beta_1 & 0 & \cdots & & 0 \\
0 & \alpha_2 & \beta_2 & & & \vdots \\
\vdots & & \ddots & \ddots & & \vdots \\
\vdots & & & \alpha_{n-1} & \beta_{n-1} & \\
\vdots & & & & \alpha_n & \\
\vdots & & & & 0 & \\
\vdots & & & & & \vdots \\
0 & & & & 0 &
\end{pmatrix},
\tag{5.44}
$$

and therefore the matrix

$$
T \;=\; B^T B = V_0^T A_{nrm} V_0,
\tag{5.45}
$$

is tridiagonal. From this remark, we deduce that V_0 can be chosen as a combination of Householder transformations as in Section 5.3.1. Let us consider the technique that reaches that goal. To avoid the computation of $A^T A$, the Householder transformations will be defined on the two sides of A. At the first step, the Householder transformation H_1^ℓ is applied on the left side of A so as to eliminate all the entries of the first column located below the diagonal. At the second step, the Householder transformation H_1^r is applied on the right side of the resulting matrix $H_1^\ell A$ so as to eliminate the entries of the first row from column 3 to column n. This transformation does not affect the zero entries of the first column that have been created in the first step. Then the process is alternatively repeated up to the last column on the two sides except the two last rows that do not need any entry elimination from the right side; when the matrix is square ($m = n$), no entries must be eliminated in the last column. The resulting bidiagonal matrix B satisfies

$$
B = H_q^\ell \cdots H_1^\ell A H_1^r \cdots H_{n-2}^r,
\tag{5.46}
$$

where $q = \min(n, m - 1)$, or equivalently $A = U_0 B V_0^T$ with $U_0 = H_1^\ell \cdots H_q^\ell$ and $V_0 = H_1^r \cdots H_{n-2}^r$.
To practically illustrate the strategy of elimination, let us consider a matrix

$A \in \mathbb{R}^{5\times 4}$. The chain of transformations is:

$$
\begin{pmatrix}
x & x & x & x \\
x & x & x & x \\
x & x & x & x \\
x & x & x & x \\
x & x & x & x
\end{pmatrix}
\xrightarrow{H_1^\ell}
\begin{pmatrix}
x & x & x & x \\
0 & x & x & x \\
0 & x & x & x \\
0 & x & x & x \\
0 & x & x & x
\end{pmatrix}
\xrightarrow{H_1^r}
\begin{pmatrix}
x & x & 0 & 0 \\
0 & x & x & x \\
0 & x & x & x \\
0 & x & x & x \\
0 & x & x & x
\end{pmatrix}
$$

$$
\xrightarrow{H_2^\ell}
\begin{pmatrix}
x & x & 0 & 0 \\
0 & x & x & x \\
0 & 0 & x & x \\
0 & 0 & x & x \\
0 & 0 & x & x
\end{pmatrix}
\xrightarrow{H_2^r}
\begin{pmatrix}
x & x & 0 & 0 \\
0 & x & x & 0 \\
0 & 0 & x & x \\
0 & 0 & x & x \\
0 & 0 & x & x
\end{pmatrix}
$$

$$
\xrightarrow{H_3^\ell}
\begin{pmatrix}
x & x & 0 & 0 \\
0 & x & x & 0 \\
0 & 0 & x & x \\
0 & 0 & 0 & x \\
0 & 0 & 0 & x
\end{pmatrix}
\xrightarrow{H_4^\ell}
\begin{pmatrix}
x & x & 0 & 0 \\
0 & x & x & 0 \\
0 & 0 & x & x \\
0 & 0 & 0 & x \\
0 & 0 & 0 & 0
\end{pmatrix}
$$

Computing the SVD of a Bidiagonal Matrix

In order to compute all the singular values of the bidiagonal matrix B, one method consists of computing the tridiagonal matrix $T = B^T B$ with the goal of computing its eigenvalues by the QR method with explicit shift or by Sturm sequences as done in Section 5.3.2. However, the most common approach is the Golub-Kahan SVD method [33]. It implicitely applies a QR step on $B^T B$ without forming the matrix $B^T B$. It uses an implicit shift that is not introduced here. The resulting procedure has a better numerical behavior since the normal matrix is not built.

5.5.2   Singular Triplets for Large Matrices

For a large and sparse matrix $A \in \mathbb{R}^{m\times n}$, it is hard or even not possible to compute its full SVD as indicated in the previous section. Only some singular triplets (σ_i, u_i, v_i) may be sought. We successively consider the two extreme situations which are the most common ones: computing the largest or the smallest singular values and their corresponding left and right singular vectors. The largest singular value σ_1 of A is equal to its 2-norm ($\|A\|_2 = \sigma_1$) whereas its smallest singular value σ_p, where $p = \min(m, n)$, corresponds to the distance of A to the set of the rank-deficient matrices in $\mathbb{R}^{m\times n}$ (for $m = n$: $\|A^{-1}\|_2 = \frac{1}{\sigma_n}$).

Computing the Largest Singular Value

For computing the largest singular value σ_1 of $A \in \mathbb{R}^{m \times n}$ where $m \geq n$, it is necessary to compute the largest eigenvalue of the matrix $A_{nrm} = A^T A$ or of the augmented matrix A_{aug} as defined in Theorem 5.1. Let us consider the matrix A_{nrm}. Its largest eigenpair (λ_1, v_1) satisfies $\lambda_1 = \sigma_1^2$ and v_1 is the right singular vector of A corresponding to σ_1. It can be computed by the Power method (see Algorithm 5.1) or by the Lanczos procedure (see Algorithm 5.6) for a faster convergence. The corresponding left singular vector u_1 is then obtained by $u_1 = \frac{1}{\|Av_1\|} A v_1$. The resulting algorithms are left to the reader as exercises (see Exercises 5.14 and 5.15).

Computing the Smallest Singular Value

For computing the smallest singular value σ_n of $A \in \mathbb{R}^{m \times n}$ where $m \geq n$, the problem is transformed into computing the largest eigenvalue μ_1 of the matrix $(A^T A)^{-1}$. This assumes that the matrix A is a full rank matrix (i.e., $\sigma_n = \frac{1}{\sqrt{\mu_1}} > 0$). In order to apply the Power method or the Lanczos procedure to the matrix $(A^T A)^{-1}$, it is necessary to solve systems with $(A^T A)$. This must be done without forming explicitly the matrix $(A^T A)$ and by considering methods that manage the sparse structure. We consider the two situations depending on the situation of A being square or not. For simplicity in the following, we omit the permutations which are the consequences of the reordering strategies used for taking advantage of the sparse pattern of the matrix.

When $m = n$, this is done by successively solving systems with matrices A and A^T. For this purpose, one may consider the LU factorization of the matrix A by using some solver dedicated to the situation of sparse matrices (see [1, 2, 3]). One may also consider iterative methods such as those that are described in the next chapter.

When $m > n$, the Cholesky factorization $A^T A = LL^T$ is obtained through the QR factorization $A = QR$, where $R \in \mathbb{R}^{n \times n}$ is upper triangular and then $L = R^T$. The QR-factorization must also be dedicated to sparse matrices (see reference [22]).

Obtaining the corresponding left singular vector u_n can be done by $u_n = \frac{1}{\|Av_n\|} A v_n$. However, the accuracy may be poor when σ_n is very small and in this case it is better to consider $u_n = \frac{1}{A^{-T} v_n} A^{-T} v_n$ (for a square matrix) or $u_n = \frac{1}{(A^+)^T v_n} (A^+)^T v_n$ with $A^{-T} = QR^{-T}$ (for a rectangular matrix). See Exercise 5.15.

5.6   Exercises

Exercise 5.1 *Eigenspace of a Jordan block.*

With the notations of Theorem 5.2, prove that the dimension of the eigenspace of J_i corresponding to λ_i is one. Prove that there is a subspace of dimension m_i which is invariant by J_i.

Exercise 5.2 *Roundoff errors when computing the eigenvalues of a Jordan block.*

In MATLAB, consider the Jordan block

$$J = \begin{pmatrix} 10 & 1 & 0 & 0 & 0 & 0 \\ 0 & 10 & 1 & 0 & 0 & 0 \\ 0 & 0 & 10 & 1 & 0 & 0 \\ 0 & 0 & 0 & 10 & 1 & 0 \\ 0 & 0 & 0 & 0 & 10 & 1 \\ 0 & 0 & 0 & 0 & 0 & 10 \end{pmatrix} \in \mathbb{R}^{6 \times 6}.$$

Build an orthogonal matrix $Q \in \mathbb{R}^{6 \times 6}$ by [Q,R]=qr(rand(6)) and define $A = QJQ^T$. What are the eigenvalues of A? Check with the command: eig. Check with other orthogonal matrices and plot all the computed eigenvalues on a common graphic. Comment.
Compute the complete diagonalization of A with eig. Check for the condition number of the matrix of eigenvectors.

Exercise 5.3 *Inverse iterations.*

Let us consider the computation of the eigenvalue of $A \in \mathbb{R}^{n \times n}$ of smallest modulus (for simplicity, it is assumed to be real). Show that this can be done by using Algorithm 5.2. Explain why, if the eigenvalue is very small it will not be computed accurately. What do you propose to cure the problem?

Exercise 5.4 *Double shift in QR.*

From the iteration (5.18) corresponding to two real shifts, prove that the relation (5.19) is true.

Exercise 5.5 *Sturm sequences and LU factorization.*

Let $T \in \mathbb{R}^{n \times n}$ be an unreduced symmetric tridiagonal matrix. What are the patterns of the factors in the LU factorization of T? When needed, express the recursions which define the two factors. Apply this result to the matrix $T - \lambda I$ and show that the quantity $q_k(\lambda)$ of the iteration (5.23) is the k-th diagonal entry of the U factor in the LU factorization of $T - \lambda I$.

Exercise 5.6 *Proof of the equality (5.30).*

For that purpose, consider a normalized vector $x \in \mathbb{R}^n$ and decompose it into $x = Vy_1 + Ry_2 + x_3$ where $V^T x_3 = R^T x_3 = 0$.
1. Prove that $\|x\|^2 = \|y_1\|^2 + \|Ry_2\|^2 + \|x_3\|^2$.
2. Then prove that

$$
\begin{aligned}
\|(RV^T + VR^T)x\|^2 &= \|Ry_1\|^2 + \|R^T Ry_2\|^2, \\
&\leq \|R\|^2 (\|y_1\|^2 + \|Ry_2\|^2), \\
&\leq \|R\|^2.
\end{aligned}
$$

3. Prove that x can be chosen such to ensure $\|(RV^T + VR^T)x\| = \|R\|$. Conclude.

Exercise 5.7 *Deflation for a conjugate pair of eigenvalues.*

Let (λ, x) be a complex eigenpair (i.e., $\lambda = \alpha + i\beta$ with the imaginary part $\beta \neq 0$) of $A \in \mathbb{R}^{n \times n}$; therefore, $(\bar{\lambda}, \bar{x})$ is a second eigenpair of A. Let $x = a + ib$ where $a, b \in \mathbb{R}^n$. Prove that $X = (a, b)$ is the basis of an invariant subspace of A. Hint: prove that $AX = XH$ with $H = \begin{pmatrix} \alpha & \beta \\ -\beta & \alpha \end{pmatrix} \in \mathbb{R}^{2 \times 2}$ and prove that $\Lambda(H) = \{\lambda, \bar{\lambda}\}$. How can you define an orthonormal basis V of the subspace span by the columns of X? Express the matrix F such that $AV = VF$.

Exercise 5.8 *Monotonicity of the sequence of Krylov subspaces.*

Let $A \in \mathbb{R}^{n \times n}$ and $v \in \mathbb{R}^n$. Let us assume that for a given $k > 1$, the dimension of the Krylov subspace $\mathcal{K}_k(A, v)$ is smaller than k. Prove that this subspace is invariant for A. Hint: consider the largest k_0 such that $\dim \mathcal{K}_{k_0}(A, v) = k_0$ and prove that $A^{k_0} v \in \mathcal{K}_{k_0}(A, v)$. Conclude that the sequence of subspaces $(\mathcal{K}_k(A, v))_{k \geq 1}$ is increasing up to some $k = k_0$ and constant for $k \geq k_0$.

Exercise 5.9 *Proof of Proposition 5.8.*

Exercise 5.10 *Invariance of a Krylov subspace.*

Prove that if the Krylov subspace $\mathcal{K}_k(A, v)$ includes an eigenvector x of A, then $\mathcal{K}_k(A, v)$ is an invariant subspace. Hint: consider the situation where $v_k^T x \neq 0$ and prove that this implies $\beta_{k+1} = 0$.

Exercise 5.11 *Deflation.*

1. Let (λ, x) be a real eigenpair of A. Prove that the spectrum of $B = (I - xx^T)A$ is $\Lambda(B) = (\Lambda(A) \backslash \{\lambda\}) \cup \{0\}$.
2. Let $V \in \mathbb{R}^{n \times p}$ an orthonormal basis of an invariant subspace of $A \in \mathbb{R}^{n \times n}$. Prove that the spectrum of $C = (I - VV^T)A$ is $\Lambda(C) = (\Lambda(A) \backslash \Lambda(H)) \cup \{0\}$, where $H = V^T AV$.

Exercise 5.12 *Lanczos without re-orthogonalization.*

Let $A \in \mathbb{R}^{n \times n}$ be a real symmetric matrix.
1. *Roundoff error effect.* Let $\tilde{V}_k$ and $\tilde{T}_k = [\tilde{\beta}_k, \tilde{\alpha}_k, \tilde{\beta}_{k+1}]$ be the matrices obtained by the Lanczos process without re-orthogonalization in floating-point

arithmetic (machine precision parameter: ϵ). We assume that

$$A\tilde{V}_k = \tilde{V}_k\tilde{T}_k + \tilde{\beta}_{k+1}\tilde{v}_{k+1}e_k^T + E_k$$

where $\|E_k\|_2 = O(\epsilon\|A\|_2)$. For any eigenpair (λ, z) of $\tilde{T}_k$ and $x = \tilde{V}_k z$ its corresponding Ritz vector, prove that

$$\left| \|Ax - \lambda x\|_2 - \tilde{\beta}_{k+1}|e_k^T z| \right| = O(\epsilon\|A\|_2). \qquad (5.47)$$

This bound allows to consider that the pair (λ, x) is an eigenpair when $\tilde{\beta}_{k+1}|e_k^T z| = O(\epsilon\|A\|_2)$.

2. *An algorithm to compute* $\Lambda(A) \cap [a, b]$, *where a and b are given*. Justify the following algorithm:

(i) Perform Lanczos without reorthogonalization to build $\tilde{V}_m$ and $\tilde{T}_m$ for a large m (e.g., $m = n$ or $m = 2n$). The basis $\tilde{V}_m$ is not stored.
(ii) Determine $\Lambda_1 = \Lambda(\tilde{T}_m) \cap [a, b] = \{\lambda_1, \cdots, \lambda_q\}$ and the corresponding eigenvectors $Z_1 = [z_1, \cdots, z_q]$.
(iii) Eliminate some spurious eigenvalues by considering the bound (5.47). The set of the remaining eigenvalues is denoted Λ_2 and the corresponding eigenvectors Z_2.
(iv) Restart the Lanczos process to compute $X_2 = \tilde{V}_m Z_2$. Reorthogonalize X_2 to eliminate copies of the duplicated eigenvectors.

In Step (iv), explain why there may be duplicated eigenpairs and how they can be eliminated by the orthogonalization.

Exercise 5.13

Prove Theorem 5.1. Give the complete description of an orthogonal matrix Q that diagonalizes A_{aug}.

Exercise 5.14 *Extremal singular values by the Power method.*

Show that the largest singular value of a matrix can be computed by the Power method. How to proceed to compute the smallest singular value?

Exercise 5.15

Do with the Lanczos procedure the same as in previous exercise.

5.7 Computer Exercises

Computer Exercise 5.1 *Computer project: Population dynamics*

Let us consider the inhabitants of a country. The whole population is partitioned into age classes. For that purpose, the following variables are defined:

(i) $x_r^{(n)}$: size of the portion of the population which is exactly r years old on the 31st of December of the year n. By assuming that no one reaches an age of 120 years, index r is such that $0 \leq r \leq 119$. The population repartition is described by the vector $x^{(n)} = \begin{pmatrix} x_0^{(n)} \\ \vdots \\ x_{119}^{(n)} \end{pmatrix}$.

(ii) $\mathcal{P}^{(n)}$: size of the whole population on the 31st of December of the year n. It can be computed by $\mathcal{P}^{(n)} = e^T x^{(n)}$ where $e \in \mathbb{R}^{120}$ is the vector with all entries equal to 1.

(iii) μ_r : rate of mortality of the age class r (i.e., ratio of the number of individuals of the class r who died during year n over $x_r^{(n-1)}$). The rate μ_r is defined for $0 \leq r \leq 119$ with $\mu_{119} = 1$. We assume that $0 < \mu_r < 1$ for $0 \leq r \leq 118$.

(iv) ρ_r : birth rate of the age class r (i.e., ratio of the number of babies who were born from one mother of the class r during year n over $x_r^{(n-1)}$). The rate ρ_r is defined for $0 \leq r \leq 119$. We assume that $0 \leq \rho_r < 1$.

1. Prove the recursion:

$$x^{(n+1)} = A x^{(n)}, \tag{5.48}$$

where A is a square matrix of order 120 defined by

$$A = \begin{pmatrix} \rho_0 & \rho_1 & \cdots & \rho_{118} & \rho_{119} \\ 1 - \mu_0 & & & & \\ & 1 - \mu_1 & & & \\ & & \ddots & & \\ & & & 1 - \mu_{118} & \end{pmatrix}$$

2. From an initial population distribution $x^{(0)}$ corresponding to a reference year $n = 0$, express the population repartition at any year n as well as the total size of the population. Give a sufficient condition $\mathbf{S_0}$ on the spectral radius of A to see that the population tends to vanish when the years pass. Give a sufficient condition $\mathbf{S_\infty}$ for it to grow indefinitely for almost every initial vector $x^{(0)}$. From $\|A\|_1$, determine a simple condition $\mathbf{S_0}$.

3. We now consider immigration and emigration. By definition vector d indicates the net of immigration: for $0 \leq r \leq 119$, the entry d_r is the net of $n+1$, between incoming and outgoing people aged r years on the 31st of December of the year n. Transform the recurrence (5.48) to express a new model. Prove that there exists a stationary population distribution $\tilde{x}$ when all the eigenvalues of A are distinct of 1. In this situation, prove that, $x^{(n+1)} - \tilde{x} = A(x^{(n)} - \tilde{x})$ and describe the population evolution with respect to the special radius of A.

Computer Exercise 5.2 *Computer Exercise: "Google" Matrices and Ordering of Web Pages*

Background on the Mathematical Model

The *World Wide Web (www)* is an unstructured set $\mathcal{W}$ which elements are "web pages." The contents of those pages are heterogeneous (all types of topics are treated) and dynamic (new pages are created and old ones edited). The main objective of a web search engine is to provide an ordering of the web pages according to a measure of the "impact" on web surfers of each web page. For that purpose, we build a mathematical model dealing with a subset of n pages $\mathcal{W}_\setminus \subset \mathcal{W}$ is by indexing the elements of $\mathcal{W}_n$, i.e.,

$$\mathcal{W}_n = \{P_1,\, P_2,\, ...,\, P_i,\, ...,\, P_n\}.$$

The common point to web pages is the existence of "links" from some pages to other ones. If page P_j has a "**hyperlink**" that points to page P_i, this is indicated by:

$$P_j \longrightarrow P_i.$$

Through the hyperlinks $\mathcal{W}_n$ can be visualized by a **graph** $\mathcal{G}(\mathcal{W}_n)$. An example of $n = 8$ pages and correspondingly its graph $\mathcal{G}(\mathcal{W}_8)$ is given in Figure 5.2. The links are as follows:
$P_1 \to P_2$; $P_2 \to P_3$; $P_3 \to P_2, P_5$; $P_4 \to P_5$; $P_5 \to P_1, P_6, P_8$; $P_6 \to P_5, P_7$; $P_7 \to P_6, P_8$; $P_8 \to P_7$.

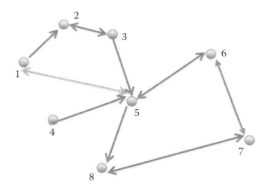

FIGURE 5.2: The graph of a set of 8 web pages

To proceed in finding the importance of a page, one starts by constructing the adjacency matrix A of $\mathcal{G}(\mathcal{W}_n)$, whereas a matrix element A_{ij} is 1 if node j has a link to node i (i.e., $P_j \to P_i$) and 0 otherwise. In the case of Figure 5.2, the adjacency matrix of links stands as follows:

$$A = A(\mathcal{G}(\mathcal{W}_8)) = \begin{pmatrix} 0 & 0 & 0 & 0 & 1 & 0 & 0 & 0 \\ 1 & 0 & 1 & 0 & 0 & 0 & 0 & 0 \\ 0 & 1 & 0 & 0 & 0 & 0 & 0 & 0 \\ 0 & 0 & 0 & 0 & 0 & 0 & 0 & 0 \\ 0 & 0 & 1 & 1 & 0 & 1 & 0 & 0 \\ 0 & 0 & 0 & 0 & 1 & 0 & 1 & 0 \\ 0 & 0 & 0 & 0 & 0 & 1 & 0 & 1 \\ 0 & 0 & 0 & 0 & 1 & 0 & 1 & 0 \end{pmatrix}$$

To find out the relevance of a page P_i in $\mathcal{W}_n$, an easy and first way is to count the number of pages k_i that point to P_i. In the above example, there are respectively 2 and 3 pages pointing to states 2 and 5. In fact (and by definition) k_i is the total number of 1's in row i of A. The highest k_{i_0} seem to indicate that page P_{i_0} is the **most cited**. However, this does not mean that P_{i_0} is the most important as it could very well be that those pages pointing to i_0 have no importance as they are not cited at all, and were just created to point to i_0.

Thus, it is essential to introduce a model which takes into account the relevance of pages P_j pointing to P_i. For that purpose let us introduce s_i the **score** of each page $P_i \in \mathcal{W}_n$. The score s_i should indicate the relevance of P_i in $\mathcal{W}_n$: $s_{i_1} > s_{i_2}$ means that P_{i_1} is more important than P_{i_2}.

Let k_j be the total number of outgoing links from page P_j to all other nodes in $\mathcal{W}_n$. Note that k_j is the sum of all 1's of column j in the adjacency matrix A of $\mathcal{W}_n$. Hence, if among k_j pages, P_j points to P_i, a proper "weight" for the link $P_j \to P_i$ is $1/k_j$ and the probability that P_j "calls" P_i is $1/k_j$. Consequently, the contribution of P_j to the score s_i of P_i is $\frac{1}{k_j}s_j$ and a linear model for the scores $\{s_i\}$ would be:

$$s_i = \sum_{P_j \to P_i} \frac{1}{k_j} s_j \tag{5.49}$$

In the example of Figure 5.2, one obtains the following 8 equations:

$$\begin{aligned} s_1 &= & & & \tfrac{1}{3}s_5 & & & \\ s_2 &= & \tfrac{1}{2}s_2 & & & & & \\ s_3 &= & s_2 & & & & & \\ s_4 &= & 0 & & & & & \\ s_5 &= & & \tfrac{1}{2}s_3 & +s_4 & & +\tfrac{1}{2}s_6 & \\ s_6 &= & & & & \tfrac{1}{3}s_5 & & +\tfrac{1}{2}s_7 \\ s_7 &= & & & & & \tfrac{1}{2}s_6 & & +s_8 \\ s_8 &= & & & & \tfrac{1}{3}s_5 & & +\tfrac{1}{2}s_7 \end{aligned} \tag{5.50}$$

This linear system (5.50) can be written in matrix form by defining a "weighted transition" matrix T transition matrix which is related to A by dividing the elements of its jth column by k_j, the total number of outgoing links from node j to all other nodes in $\mathcal{W}_n$. In the above example, the "weighted" matrix T of $\mathcal{W}_8$ is given by:

$$T = T(\mathcal{G}(\mathcal{W}_8)) = \begin{pmatrix} 0 & 0 & 0 & 0 & 1/3 & 0 & 0 & 0 \\ 1 & 0 & 1/2 & 0 & 0 & 0 & 0 & 0 \\ 0 & 1 & 0 & 0 & 0 & 0 & 0 & 0 \\ 0 & 0 & 0 & 0 & 0 & 0 & 0 & 0 \\ 0 & 0 & 1/2 & 1 & 0 & 1/2 & 0 & 0 \\ 0 & 0 & 0 & 0 & 1/3 & 0 & 1/2 & 0 \\ 0 & 0 & 0 & 0 & 0 & 1/2 & 0 & 1 \\ 0 & 0 & 0 & 0 & 1/3 & 0 & 1/2 & 0 \end{pmatrix}$$

and in matrix form, the linear system (5.50) can be written as:

$$s = Ts \qquad (5.51)$$

i.e., for this problem to have a solution, the matrix T should have the eigenvalue 1 with s a corresponding eigenvector.

Remark 5.7 "Dangling" Pages *The case of a column j of A that has all zero elements corresponds to a "dangling node" ($k_j = 0$: no outgoing link from P_j).*

In such case we modify column j of a dangling node in the above defined T by letting each of its elements equal to $p = \frac{1}{n}$. Such p represents the probability that node j makes a request to any page in $\mathcal{W}_n$. The resulting transformed matrix T^d is thus obtained from T as follows:

$$T^d_{.j} = T_{.j} \text{ if } T_{.j} \neq 0 \text{ and } T^d_{.j} = p, \text{ otherwise.}$$

The matrix T^d has non-negative entries and the sum of the entries in each column is one. It is therefore a stochastic matrix. Stochastic matrices have several properties that will prove useful to us. For instance, stochastic matrices always have stationary vectors. Note that T^d can also be obtained from T by considering the matrix C whose entries are all zero except for the columns corresponding to dangling nodes where each entry is $1/n$. Then $T^d = T + C$. More generally, the elements of a Google matrix $G = G(\mathcal{W}_n)$ associated with $\mathcal{W}_n$ are given by:

$$G_{ij}(\delta) = \delta T^d_{ij} + (1 - \delta)p \qquad (5.52)$$

where the coefficient δ is a damping factor, $(0 < \delta < 1)$ with $(1 - \delta)$ giving an additional probability to direct randomly in $\mathcal{W}_n$ from page P_j to page P_i. Note the following:

1. In the case of a dangling node j, $T^d_{ij} = G_{ij} = p, \forall i$.
2. The sum of all non-negative elements of each of its columns is equal to unity.

In matrix form (5.52) becomes:

$$G = \delta T^d + (1 - \delta)E, \tag{5.53}$$

where E is the $n \times n$ matrix which elements are all $p = \frac{1}{n}$.

For $0 < \delta < 1$ the matrix G is a Markov chain type matrix. Using the Perron-Frobenius theory for Markov chains type matrices its spectrum $\Lambda(G) \subset \{z \in \mathbb{C} \,|\, |z| \leq 1\}$ and furthermore there is only *one maximal eigenvalue* $\lambda = 1$ with algebraic and geometric multiplicity equal to 1 and a corresponding eigenvector s which has non-negative elements s_i.

Use of Power Method to Compute the PageRank Vector

On the basis of (5.51), the vector of page scores s of $\mathcal{W}_n$ satisfies:

$$s = Gs = \delta T^d s + (1 - \delta)Es \tag{5.54}$$

These components ordered by decreasing values give the *PageRank vector* according to the matrix $G(\delta)$.

Starting with an arbitrary vector $s^{(0)}$ with positive components, one considers the iterative scheme:

$$s^{(k+1)} = Gs^{(k)}, \; k \geq 0. \tag{5.55}$$

which is stopped whenever we reach k_ϵ such that:

$$\frac{||s^{(k_\epsilon)} - s^{(k_\epsilon - 1)}||_\infty}{||s^{(k_\epsilon)}||_\infty} < \epsilon.$$

Computationally, note that the Gaxpy operation in (5.55) proceeds as follows. For a vector $x \in \mathbb{R}^n$:

$$Gx = \delta T^d x + (1 - \delta)Ex.$$

Since T^d is a sparse matrix, the number of operations to perform $T^d x$ is $O(k_m n)$ where $k_m = \max_{1 \leq i \leq n} k_i$ where k_i is the number of non-zero elements in the ith row of T^d. On the other hand $(1 - \delta)Ex = p(1 - \delta)\sum_{i=1}^n x_i \mathbf{1}$, where $\mathbf{1}$ is the vector that has all its n components equal to 1.

This implies that the total number of flops to perform one iteration of (5.55) is $O(k_m n)$.

The PageRank Algorithm

Consider Algorithm 5.7 which implements (5.55), then answer the following **questions**:

1. In Step 3 of Algorithm 5.7, fill the statement `y=Gx` without forming the Google matrix G. Give the number of flops to execute the algorithm in terms of n, the number of iterations n_{it} and $k_m = \max_{1 \leq i \leq n} k_i$ where k_i is the number of non-zero elements in the ith row of $T^{\bar{d}}$.

2. Apply Algorithm 5.7 to the examples drawn in Figure 5.2 by taking the following values of $\delta = 0.2, 0.3, 0.4, 0.5, 0.6, 0.7, 0.8, 0.9$ and modify the program so as to find out which δ's provide a faster convergence. Is the value of $\delta = 0.85$ chosen by Brin and Page [14] appropriate?

3. Same question as 1 applied to the network drawn in Figure 5.3 (used in [7]).

4. Same question as 1 applied to the network drawn in Figure 5.4 (used in [6]).

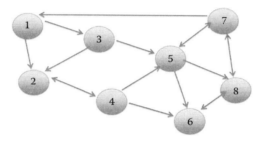

FIGURE 5.3: A graph for a network of 8 web pages

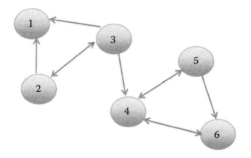

FIGURE 5.4: A graph for a network of 6 web pages

Algorithm 5.7 The Power Method to Find the PageRank of a Network

---

```
function [x,Ix]=MyPageRank(n,I,J,c,delta,tol)
% Input: n is the number of pages of the network
%   Table I, J, c  describes the hyperlinks:
%            I(1)=i; I(1)=j; c(1)=1; 1<=l<=k; 1<=i<=j<=n
% Output: The Pagerank vector x with 0<=x(i)<=10
%   The vector Ix of page importances according to decreasing order
% Step 1: Form the matrices T and Td
e=ones(n,1);T=sparse(I,J,c,n,n);sT=sum(T);
for j=1:n
     Jj= J==j;
     if sT(j)==0
         c(Jj)=1/n;
     else
         Jj=find(J==j);c(Jj)=(1/sT(j))*c(Jj);
     end
end
Td=sparse(I,J,c,n,n);
% Form the Google matrix G=delta*Td+(1-delta)/n
% Initialize the iteration
x=rand(n,1); y=zeros(n,1);r= y - x ;%norm(x,'inf')<1
% Start the iteration
while norm(r,inf)/norm(x,inf) > tol
       %===> |Fill in Statement for y=Gx|
    x = y/norm(y,inf) ;r = y - x ;
end
x=fix(1000*x)/100;[x,Ix]=sort(x,'descend');
```

Chapter 6

Iterative Methods for Systems of Linear Equations

In this chapter we discuss iterative methods for solving

$$Ax = b, \tag{6.1}$$

where $x \in \mathbb{R}^n$, $A \in \mathbb{R}^{n \times n}$ and $b \in \mathbb{R}^n$ are given. We assume that A is non-singular, so that problem (6.1) has a unique solution

$$x = A^{-1}b. \tag{6.2}$$

Starting from an initial guess $x^{(0)}$, iterative methods generate a sequence $\{x^{(0)}, x^{(1)}, \ldots, x^{(k)}, \ldots, \}$ which converges towards $A^{-1}b$. We define the residuals as

$$r^{(k)} = b - Ax^{(k)}, \tag{6.3}$$

and the errors as

$$e^{(k)} = x - x^{(k)} = A^{-1}r^{(k)}. \tag{6.4}$$

Thus the sequence of residuals and the sequence of errors converge towards 0 when the iterative method converges. The rate of convergence should be fast enough so that iterations can stop when an accurate approximation is found. Each iteration requires BLAS 2 GAXPY's operations with a complexity of $O(n^2)$ flops for dense matrices and $O(n)$ flops for sparse matrices. In practice, the system must be preconditioned with another matrix and each iteration requires solving a system with this matrix. The objectives of preconditioning are twofold: reduce the number of iterations and solve an easy system, with a linear $O(n)$ complexity for sparse matrices. Moreover, memory requirements are also linear in $O(n)$ floating-point numbers. When convergence is fast enough, iterative methods are competitive with direct methods, which require much more memory and have a higher complexity.

There are several references on iterative methods to solve a system of linear equations, for example Kelley [42], Saad [58], Ciarlet [20], Meurant [47] and Dongarra et al. [26].

In this chapter, we discuss only two classes of iterative methods: stationary methods and Krylov projection methods. We describe some well-known methods and refer to the literature for the others. We do not discuss other iterative methods, such as multigrid methods or domain decomposition methods.

6.1 Stationary Methods

Stationary methods are classical iterative methods based on a splitting of the matrix. In most cases, they converge slowly so that they are no longer used alone, but rather used as smoothers in multigrid methods.

6.1.1 Splitting

Stationary methods are based on a decomposition

$$A = M - N, \tag{6.5}$$

where M is a non-singular matrix. Iterations are based on a fixed point method defined by

$$Mx^{(k+1)} = Nx^{(k)} + b. \tag{6.6}$$

Clearly, if the method converges, then the limit is the solution of (6.1). We get $Mx^{(k+1)} = (M - A)x^{(k)} + b = Mx^{(k)} + r^{(k)}$ so that

$$x^{(k+1)} = x^{(k)} + M^{-1}r^{(k)}.$$

We also get $Mx^{(k+1)} = Nx^{(k)} + (M - N)x$ thus $M(x - x^{(k+1)}) = N(x - x^{(k)})$ and

$$e^{(k+1)} = (M^{-1}N)e^{(k)},$$

finally $e^{(k)} = (M^{-1}N)^k e^{(0)}$.

Thus the error converges towards 0 when $\|M^{-1}N\| < 1$ for some matrix norm. More precisely, the following theorem can be proved.

Theorem 6.1 *Let ρ be the spectral radius of the matrix $(M^{-1}N)$, coming from the splitting $A = M - N$ of the non-singular matrix A. The method (6.6) is convergent for any initial vector $x^{(0)}$ if and only if $\rho < 1$.*

6.1.2 Classical Stationary Methods

The classical examples are Jacobi, Gauss-Seidel, SOR and SSOR methods. Let $A = D - E - F$ be the splitting of A in a diagonal part, lower triangular part, upper triangular part.

- Jacobi method is defined by $M = D$, with the iterations

$$Dx^{(k+1)} = (E + F)x^{(k)} + b.$$

- Gauss-Seidel method is defined by $M = D - E$, with the iterations
$$(D - E)x^{(k+1)} = Fx^{(k)} + b.$$

- SOR method is defined by $M = D - \omega E$ and solves $\omega Ax = \omega b$, where ω is a parameter chosen to speed up convergence. Iterations are
$$(D - \omega E)x^{(k+1)} = (\omega F + (1 - \omega)D)x^{(k)} + \omega b.$$

- SSOR method uses first an SOR iteration followed by one other SOR iteration in reverse order. Thus
$$M = \frac{1}{\omega(2 - \omega)}(D - \omega E)D^{-1}(D - \omega F).$$

Convergence can be proved for certain classes of matrices, see for example [33] and [47]. We recall here some important results.

Definition 6.1 *A matrix $A = (a_{ij})$ of order n is strictly diagonally dominant if*
$$|a_{jj}| > \sum_{i=1, i \neq j}^{n} |a_{ij}|, \; j = 1, \ldots, n$$

Theorem 6.2 *If A is strictly diagonally dominant, Jacobi and Gauss-Seidel methods converge for any $x^{(0)}$.*

Theorem 6.3 *Let A be a symmetric matrix with positive diagonal elements and let $0 < \omega < 2$. SOR method converges for any $x^{(0)}$ if and only if A is s.p.d.*

Algorithm 6.1 Relaxation Iterative Method $(0 < \omega < 2)$

```
function [x,r,k]=RM(A,b,Omega,kmax,eps1)
% kmax maximum number of iterations;
% eps1 given tolerance;
% Omega: relaxation parameter
% stopping criteria: norm(r)/norm(b)<=eps1 or k>kmax
% Output: x the kth iterate, r the kth residual
%         k the number of iterations
% Initial guess is x=0; r=b
n=length(b);r=b;x=zeros(n,1);nb=norm(b);
E=tril(A,-1);D=diag(diag(A));M=(1/Omega)*D-E;
k=1;mr=1;mx=1;e=1;
while e>eps1 & k<=kmax
        dx=M\r;
        x=x+dx;% saxpy operation
        r=r-A*dx;% GAXPY operation
        e=norm(r)/nb;
        k=k+1;
end
```

6.2   Krylov Methods

Krylov methods are iterative methods using simple vector operations and matrix-vector products. They are described in many papers, for example [15], [27] and [61].

6.2.1   Krylov Properties

Krylov methods are strongly related to polynomial properties.
Let $p_A(\lambda) = \det(A - \lambda I)$, be the characteristic polynomial associated with a matrix $A \in \mathbb{R}^{n \times n}$. From Cayley-Hamilton theorem one has:

$$p_A(A) = 0,$$

where $p_A(\lambda) = \det(A) + \sum_{i=1}^{n} c_i \lambda^i$, with real coefficients $c_i, i = 1, .., n$.

Since A is non-singular, $\det(A) \neq 0$ and $\forall\, y \in \mathbb{R}^n$

$$\det(A)y = -\sum_{i=1}^{n} c_i A^i y, \tag{6.7}$$

$$A^{-1}y = -\frac{1}{\det(A)} \sum_{i=0}^{n-1} c_{i+1} A^i y. \tag{6.8}$$

Let $x = A^{-1}b$ the solution to $Ax = b$, $x^{(0)}$ any vector, $e^{(0)} = x - x^{(0)}$ and $r^{(0)} = b - Ax^{(0)} = Ae^{(0)}$. Letting $y = r^{(0)}$ in the above equation, we get

$$e^{(0)} = A^{-1}r^{(0)} = -\frac{1}{\det(A)} \sum_{i=0}^{n-1} c_{i+1} A^i r^{(0)}.$$

We conclude that:

$$A^{-1}b \in x^{(0)} + \mathcal{K}_n(A, r^{(0)}),$$

where:

$$\mathcal{K}_n(A, r^{(0)}) = span\{r^{(0)}, Ar^{(0)}, A^2 r^{(0)}, ..., A^{n-1}r^{(0)}\}$$

is the nth Krylov subspace associated to A and $r^{(0)}$. More precisely, we have the following useful result.

Proposition 6.1 *Let $1 \leq p \leq n$ be the smallest integer such that $A^{-1}b \in x^{(0)} + \mathcal{K}_p(A, r^{(0)})$. Then*

$$\begin{cases} \mathcal{K}_k(A, r^{(0)}) = \mathcal{K}_p(A, r^{(0)}), k \geq p, \\ dim(\mathcal{K}_k(A, r^{(0)})) = k, k \leq p. \end{cases}$$

Converserly, if $A^{-1}b \in x^{(0)} + \mathcal{K}_k(A, r^{(0)})$ then $\mathcal{K}_{k+1}(A, r^{(0)}) = \mathcal{K}_k(A, r^{(0)})$ and $k \geq p$.

Proof. First, let us remark that $A^{-1}b \in x^{(0)} + \mathcal{K}_k(A, r^{(0)})$ is equivalent to $A^{-1}r^{(0)} \in \mathcal{K}_k(A, r^{(0)})$.

Now, let us assume that $A^{-1}r^{(0)} \in \mathcal{K}_p(A, r^{(0)})$ and $A^{-1}r^{(0)} \notin \mathcal{K}_{p-1}(A, r^{(0)})$. Then $A^{-1}r^{(0)} = \sum_{i=0}^{p-1} \alpha_i A^i r^{(0)}$ with $\alpha_{p-1} \neq 0$. By multiplying by A and by dividing by α_{p-1}, it comes that $A^p r^{(0)} \in \mathcal{K}_p(A, r^{(0)})$, thus $\mathcal{K}_{p+1}(A, r^{(0)}) = \mathcal{K}_p(A, r^{(0)})$ and by recurrence, $\mathcal{K}_k(A, r^{(0)}) = \mathcal{K}_p(A, r^{(0)}), k \geq p$.

Now, let us assume that $dim(\mathcal{K}_p(A, r^{(0)})) \leq p-1$; then there exists a nontrivial linear combination such that $\sum_{i=0}^{p-1} \alpha_i A^i r^{(0)} = 0$; let $0 \leq k \leq p-2$ the largest integer such that $\alpha_k \neq 0$. Then by multiplying by $A^{-(k+1)}$ and dividing by α_k, it comes that $A^{-1}r^{(0)} \in \mathcal{K}_{p-(k+1)}(A, r^{(0)})$, so that p is not the smallest integer verifying this property. Therefore $dim(\mathcal{K}_p(A, r^{(0)})) = p$ and by induction, it is true also for $k \leq p$. ∎

This property motivates the subspace condition of Krylov methods.

6.2.2   Subspace Condition

Iterative Krylov methods define sequences $x^{(0)}, x^{(1)}, ..., x^{(k)}, ..$, such that

$$x^{(k)} \in x^{(0)} + \mathcal{K}_k(A, r^{(0)}), k \geq 1.$$

As previously, let us define the residuals as $r^{(k)} = b - Ax^{(k)}$ and the errors as $e^{(k)} = x - x^{(k)} = A^{-1}r^{(k)}$. Clearly,

$$r^{(k)} \in \mathcal{K}_{k+1}(A, r^{(0)}).$$

It should be noted that iterative Krylov methods can converge in at most p iterations, where $p \leq n$ is defined in Proposition 6.1, since the solution belongs to $\mathcal{K}_p(A, r^{(0)})$.

Also, as soon as $r^{(k)} = 0$, the solution is found and the algorithm can stop. Thus, it is natural to assume that $r^{(j)} \neq 0, j \leq k$ when defining the next iterate.

A Krylov subspace is strongly related to polynomials. Indeed, the subspace condition can be equivalently written in a polynomial form:

$$x^{(k)} = x^{(0)} + s_{k-1}(A)r^{(0)}$$

Thus, the residuals of Krylov methods can be expressed as polynomials of A applied to $r^{(0)}$. This will be useful for convergence analysis.

Proposition 6.2 *The residual and the error satisfy*

$$r^{(k)} = q_k(A)r^{(0)}, \ e^{(k)} = q_k(A)e^{(0)}, \tag{6.9}$$

where q_k is a polynomial of degree k such that $q_k(0) = 1$.

Proof.

$$r^{(k)} = b - A(x^{(0)} + s_{k-1}(A)r^{(0)}) = r^{(0)} - As_{k-1}(A)r^{(0)} = q_k(A)r^{(0)},$$

where $q_k(X) = 1 - Xs_{k-1}(X)$.

$$e^{(k)} = A^{-1}r^{(k)} = A^{-1}q_k(A)r^{(0)} = q_k(A)A^{-1}r^{(0)} = q_k(A)e^{(0)}.$$

∎

In order to define the iterates, the subspace condition can be written at least in two different ways.

First, it is equivalent to

$$x^{(k+1)} = x^{(k)} + \alpha_k p^{(k)}, k \geq 0, \alpha_k \in \mathbb{R}, p^{(k)} \in \mathcal{K}_{k+1}(A, r^{(0)}), \qquad (6.10)$$

giving a recurrence relation between two consecutive iterates. The residuals satisfy also a recurrence

$$r^{(k+1)} = r^{(k)} - \alpha_k A p^{(k)}. \qquad (6.11)$$

The method must define the coefficient α_k and the descent direction $p^{(k)}$. Second, the subspace condition is equivalent to

$$x^{(k)} = x^{(0)} + V_k y^{(k)}, y^{(k)} \in \mathbb{R}^k, \qquad (6.12)$$

where V_k is a basis of $\mathcal{K}_k(A, r^{(0)})$. The residuals satisfy

$$r^{(k)} = r^{(0)} - AV_k y^{(k)}. \qquad (6.13)$$

The method must define the basis V_k, and the vector $y^{(k)}$. It also requires computing AV_k.

The definition of the descent direction or the Krylov basis can be done through a minimization property.

6.2.3   Minimization Property for spd Matrices

If A is spd, the unique solution to (6.1) can be considered as a solution of an optimization problem. Let us define the quadratic function

$$\phi(y) = \frac{1}{2}y^T Ay - y^T b, \forall\, y \in \mathbb{R}^n.$$

The gradient of ϕ satisfies

$$\nabla\phi(y) = -b + Ay.$$

Thus, if ϕ has a minimum value at y, then $\nabla\phi(y) = 0$ and $y = x$, the solution of (6.1). The following proposition gives a sufficient condition to satisfy this minimum property.

Proposition 6.3 *If A is symmetric, the unique solution x of (6.1) satisfies*

$$\phi(y) - \phi(x) = \frac{1}{2}(y - x)^T A(y - x), \forall\, y \in \mathbb{R}^n. \qquad (6.14)$$

Furthermore, if A is spd, then x satisfies

$$\phi(x) = \min\{\phi(y), y \in \mathbb{R}^n\}.$$

Proof. The solution x and $y \in \mathbb{R}^n$ satisfy the identity:

$$\phi(y) - \phi(x) = \frac{1}{2}[y^T A y - x^T A x] - (y - x)^T b.$$

By symmetry of A, we get $y^T A y - x^T A x = (y - x)^T A(y + x)$; moreover, since x is solution of (6.1), $(y - x)^T b = (y - x)^T A x$. Thus

$$\phi(y) - \phi(x) = \frac{1}{2}(y - x)^T A(y - x).$$

Now, if A is spd, then $(y - x)^T A(y - x) = \|y - x\|_A^2$, yielding $\phi(y) \geq \phi(x)$. Moreover, $\|y - x\|_A = 0 \Leftrightarrow y = x$ thus x is the unique vector which minimizes the function ϕ. ∎

When A is spd, this minimization property is used to define the Krylov methods called steepest descent and conjugate gradient.

Let E be a subspace of $\mathbb{R}^n$. To minimize the quadratic function ϕ in this subspace is equivalent to an orthogonality condition.

Proposition 6.4 *Let A be an spd matrix, E a subspace of $\mathbb{R}^n$, and $y \in \mathbb{R}^n$. The following conditions are equivalent:*

$$\phi(y) \quad = \quad \min_{z \in E} \phi(z), \tag{6.15}$$

$$\|y - x\|_A \quad = \quad \min_{z \in E} \|z - x\|_A, \tag{6.16}$$

$$y - x \perp_A E, \tag{6.17}$$

$$b - Ay \perp E. \tag{6.18}$$

Proof. We already saw that $\phi(y) = \phi(x) + \frac{1}{2}\|y - x\|_A^2$, thus the two first conditions are equivalent.

Now, to minimize the $A-$ distance of x to the subspace E is equivalent to find $y \in E$ such that $y - x$ is $A-$ orthogonal to E. Thus the third condition is equivalent.

Since $\forall z \in E, z^T A(y - x) = z^T(Ay - b)$, the fourth condition is equivalent. ∎

6.2.4   Minimization Property for General Matrices

In the general case, x can always be thought of as a least squares solution.

Proposition 6.5 *The solution x is the unique solution of*

$$\|b - Ax\| = \min\{\|b - Ay\|, y \in \mathbb{R}^n\},$$

$\|.\|$ *being any norm in $\mathbb{R}^n$.*

This least squares problem is used to define the Krylov method GMRES.

6.3 Method of Steepest Descent for spd Matrices

Let A be an spd matrix.

The steepest descent method uses the recurrence (6.10) and minimizes the function ϕ over the one-dimensional subspace $E = \{y = x^{(k)} + \alpha p^{(k)}\}$.

Using Proposition 6.4, this is equivalent to find $x^{(k+1)}$ such that the residual $r^{(k+1)}$ is orthogonal to E. Since $p^{(k)}$ is a basis of E, it is equivalent to say that $r^{(k+1)}$ is orthogonal to $p^{(k)}$.

This orthogonality condition is satisfied by

$$\alpha_k = \frac{(r^{(k)})^T p^{(k)}}{(p^{(k)})^T A p^{(k)}}.$$

It can be noted that α_k is well defined since A is spd.

Only the real parameter α_k is defined by this condition. The descent direction is chosen here equal to the opposite of the gradient of ϕ, that is to say the residual $r^{(k)}$, as long as $r^{(k)} \neq 0$. It is a steepest descent direction in the sense that $\nabla \phi(x^{(k)})^T p = -(r^{(k)})^T p$ with $\|p\| = 1$ is minimal for $p^{(k)} = r^{(k)}/\|r^{(k)}\|$.

This proves the following proposition:

Proposition 6.6 *As long as $r^{(k)} \neq 0$, the steepest descent iterative method defines the sequence $\{x^k\}, k \geq 0$ by:*

$$
\begin{align}
x^{(0)} & \quad given, & (6.19)\\
r^{(0)} &= b - Ax^{(0)}, & (6.20)\\
\alpha_k &= \frac{(r^{(k)})^T r^{(k)}}{(r^{(k)})^T A r^{(k)}}, & (6.21)\\
x^{(k+1)} &= x^{(k)} + \alpha_k r^{(k)}, & (6.22)\\
r^{(k+1)} &= r^{(k)} - \alpha_k A r^{(k)}. & (6.23)
\end{align}
$$

It is such that $\phi(x^{(k+1)}) = \min_\alpha(\phi(x^{(k)} + \alpha r^{(k)})$.

The procedure for the steepest descent method is depicted in Algorithm 6.2.

Algorithm 6.2 Algorithm of Steepest Descent

```
% x0,b given; eps1 given tolerance
% kmax  maximum number of iterations,
x=x0;r=b-A*x0;k=0;nb=norm(b);e=1;
while e>eps1 & k<=kmax
   k=k+1;nr=r'*r;q=A*r;nar=r'*q;alpha=nr/nar;
   x=x+alpha*r;r=r-alpha*q;
   e=sqrt(nr)/nb;
end
```

6.3.1 Convergence Properties of the Steepest Descent Method

We recall that the matrix A is spd.
Using the minimization property, we first prove that the convergence is monotonous.

Proposition 6.7 *We assume that $r^{(j)} \neq 0, j \leq k$. In the steepest descent method, the errors in the $A-$norm decrease:*

$$\|e^{(k+1)}\|_A \leq \|e^{(k)}\|_A.$$

Proof. We have $\phi(x^{(k)}) = \phi(x) + \frac{1}{2}\|e^{(k)}\|_A^2$ and $\phi(x^{(k+1)}) \leq \phi(x^{(k)})$ thus $\|e^{(k+1)}\|_A \leq \|e^{(k)}\|_A.$ ∎

Now, using the subspace condition, we can prove a stronger result, with a strictly monotonous convergence.

Proposition 6.8 *In the steepest descent method, the errors in the $A-$norm strictly decrease:*

$$\|e^{(k+1)}\|_A \leq (1 - \frac{1}{\kappa(A)})^{1/2}\|e^{(k)}\|_A, \tag{6.24}$$

where $\kappa(A) = \|A\|\|A^{-1}\|$ is the condition number of the matrix A.

Proof.

$$\begin{aligned}
\|e^{(k+1)}\|_A^2 &= (r^{(k+1)})^T A^{-1} r^{(k+1)} \\
&= (r^{(k)} - \alpha_k Ar^{(k)})^T A^{-1}(r^{(k)} - \alpha_k Ar^{(k)}).
\end{aligned}$$

Using $\alpha_k = \frac{(r^{(k)})^T r^{(k)}}{(r^{(k)})^T Ar^{(k)}}$ we get

$$\|e^{(k+1)}\|_A^2 = \|e^{(k)}\|_A^2 - \alpha_k \|r^{(k)}\|^2.$$

Now, $(r^{(k)})^T Ar^{(k)} \leq \|A\|\|r^{(k)}\|^2$, thus $\alpha_k \geq \frac{1}{\|A\|}$.
Also, $\|e^{(k)}\|_A^2 \leq \|A\|^{-1}\|r^{(k)}\|^2$, thus $\|r^{(k)}\|^2 \geq \frac{1}{\|A\|^{-1}}\|e^{(k)}\|_A^2$.
Putting things together, we get the result wanted. ∎

6.3.2 Preconditioned Steepest Descent Algorithm

To improve convergence, we precondition the matrix A in writing an equivalent system to (6.1):
$$CAC^T(C^{-T}x) = Cb,$$

where C is invertible. Note that CAC^T is also spd The choice of C should be such that $\kappa(CAC^T) << \kappa(A)$. Algorithm 6.2 can then be rewritten using the spd matrix M defined by
$$M^{-1} = C^T C.$$

By introducing the preconditioned residual vector $z = M^{-1}r$, we get the following proposition.

Proposition 6.9 *As long as $r^{(k)} \neq 0$, the preconditioned steepest descent method defines the sequence $\{x^k\}, k \geq 0$ by*

$$x^{(0)} \quad given, \tag{6.25}$$

$$r^{(0)} = b - Ax^{(0)}, \tag{6.26}$$

$$z^{(0)} = M^{-1}r^{(0)}, \tag{6.27}$$

$$\alpha_k = \frac{(z^{(k)})^T r^{(k)}}{(z^{(k)})^T A z^{(k)}}, \tag{6.28}$$

$$x^{(k+1)} = x^{(k)} + \alpha_k z^{(k)}, \tag{6.29}$$

$$r^{(k+1)} = r^{(k)} - \alpha_k A z^{(k)}, \tag{6.30}$$

$$z^{(k+1)} = M^{-1}r^{(k+1)}. \tag{6.31}$$

The procedure for the preconditioned steepest descent method is depicted in Algorithm 6.3.

Algorithm 6.3 Preconditioned Steepest Descent

```
% x0 (initial condition), b given;
% M is the preconditioner
% eps1 is the tolerance;
% kmax is the maximum number of iterations
x=x0;r=b-A*x0;z=M\r;
k=0;e=1;nb=norm(b);
while e>eps1 & k<=kmax
    k=k+1;
    z=M\r;q=A*z;
    nr=z'*r;nar=z'*q;alpha=nr/nar;
    x=x+alpha*z;r=r-alpha*q;
    e=norm(r)/nb;
end
```

6.4   Conjugate Gradient Method (CG) for spd Matrices

Let A be an spd matrix. The Conjugate Gradient method, noted CG, is a Krylov method which in general converges faster than the steepest descent. It still uses the recurrence (6.10) but it minimizes the function ϕ over the entire Krylov subspace.

In other words, $\phi(x^{(k)}) = \min_{y \in \{x^{(0)} + \mathcal{K}_k(A, r^{(0)})\}} \phi(y)$, or equivalently

$$\|e^{(k)}\|_A = \min_{y \in \{x^{(0)} + \mathcal{K}_k(A, r^{(0)})\}} \|y - x\|_A. \tag{6.32}$$

This minimization property is equivalent to an orthogonality condition.

Proposition 6.10 *In the Conjugate Gradient method, the residuals satisfy*

$$(r^{(k)})^T v = 0, \forall v \in \mathcal{K}_k(A, r^{(0)}). \tag{6.33}$$

Proof. We simply apply Proposition 6.4. ∎
This condition defines both the scalar α_k and the descent direction $p^{(k)}$.

6.4.1   Krylov Basis Properties

The orthogonality conditions imply that the residuals form an orthogonal basis of the Krylov subspace, provided they are non-zero. If a residual is zero, then the solution is found and the algorithm can stop.

Proposition 6.11 *Let us assume that the residuals are non-zero. The sequence of residuals $\{r^{(0)}, \ldots, r^{(k-1)}\}$ obtained from the Conjugate Gradient method is an orthogonal basis of the Krylov subspace $\mathcal{K}_k(A, r^{(0)})$.*

Proof. Since the k residuals are mutually orthogonal and non-zero, they are linearly independent vectors. Moreover, they are in the subspace $\mathcal{K}_k(A, r^{(0)})$, which is of dimension at most k, thus the residuals form a basis. ∎
Before proving a similar result about the descent directions, we first prove that they are non-zero.

Lemma 6.1 *If $r^{(k)} \neq 0$, then $p^{(k)} \neq 0$ and $\alpha_k \neq 0$ and $x^{(k+1)} \neq x^{(k)}$.*

Proof. If $p^{(k)} = 0$ or $\alpha_k = 0$ then, since $r^{(k+1)} = r^{(k)} - \alpha_k A p^{(k)}$, it implies $r^{(k+1)} = r^{(k)}$. But $r^{(k+1)}$ is orthogonal to $r^{(k)}$, thus $r^{(k)} = 0$. Since A is non-singular, the first part is proved.
Now, if $p^{(k)} \neq 0$ and $\alpha_k \neq 0$, $x^{(k+1)} \neq x^{(k)}$ since $x^{(k+1)} = x^{(k)} + \alpha_k p^{(k)}$. ∎
Because the matrix is symmetric, the orthogonality property of the residuals implies a $A-$ orthogonality of the descent directions.

Proposition 6.12 *As long as the residuals are non-zero, the descent directions $\{p^{(0)}, \ldots, p^{(k-1)}\}$ obtained from the Conjugate Gradient method is $A-$ orthogonal basis of the Krylov subspace $\mathcal{K}_k(A, r^{(0)})$.*

Proof. Since $r^{(k+1)} = r^{(k)} - \alpha_k A p^{(k)}$ and $\alpha_k \neq 0$, the orthogonality property of the residuals implies that $A p^{(k)}$ is also orthogonal to the Krylov subspace $\mathcal{K}_k(A, r^{(0)})$.
Since $p^{(j)} \in \mathcal{K}_{j+1}(A, r^{(0)})$, we get $(A p^{(k)})^T p^{(j)} = 0, j \leq k - 1$, and by symmetry of A, $(p^{(k)})^T A p^{(j)} = 0, j \neq k$.

Since the residuals are non-zero, the descent directions are also non-zero. Because A is spd, $(p^{(k)})^T A p^{(k)} \neq 0$, so that the descent directions form a $A-$ orthogonal basis of the Krylov subspace.   ∎

Since the descent directions form a basis, we can prove the following lemma.

Lemma 6.2 *As long as the residuals are non-zero, $p^{(k)} \in \mathcal{K}_{k+1}(A, r^{(0)})$ but $p^{(k)} \notin \mathcal{K}_k(A, r^{(0)})$.*

Proof. The first part is in the definition of the subspace condition.
Assume that $p^{(k)} \in \mathcal{K}_k(A, r^{(0)})$, then, thanks to the $A-$ orthogonality, $(p^{(k)})^T A p^{(k)} = 0$. Because A is spd, this implies that $p^{(k)} = 0$. But $p^{(k)} \neq 0$ thus $p^{(k)} \notin \mathcal{K}_k(A, r^{(0)})$.   ∎

We are now able to define the descent direction $p^{(k)}$.

Proposition 6.13

$$p^{(k+1)} = r^{(k+1)} + \beta_k p^{(k)}, k \geq 0$$

Proof. For $k \geq 1$, we can write $p^{(k)} = r^{(k)} + v$ with $v \in \mathcal{K}_k(A, r^{(0)})$, where we have chosen the non-zero coefficient of $r^{(k)}$ equal to 1. For $k = 0$, we also choose $p^{(0)} = r^{(0)}$.

Since the descent directions form a basis of the Krylov subspace, we can finally write

$$p^{(k)} = r^{(k)} + \sum_{j=0}^{k-1} \beta_j p^{(j)}.$$

Thus $(p^{(j)})^T A p^{(k)} = (p^{(j)})^T A r^{(k)} + \beta_j (p^{(j)})^T A p^{(j)}$. For $j \leq k - 2$, we have $A p^{(j)} \in \mathcal{K}_k(A, r^{(0)})$ thus $(p^{(j)})^T A r^{(k)} = 0$. Since $(p^{(j)})^T A p^{(k)} = 0$ and $(p^{(j)})^T A p^{(j)} \neq 0$, we conclude that $\beta_j = 0$ and

$$p^{(k)} = r^{(k)} + \beta_{k-1} p^{(k-1)}, k \geq 1.$$

∎

Let us now define the scalars α_k and β_k.

Proposition 6.14

$$\alpha_k = \frac{(r^{(k)})^T p^{(k)}}{(p^{(k)})^T A p^{(k)}}.$$

$$\beta_k = -\frac{(r^{(k+1)})^T A p^{(k)}}{(p^{(k)})^T A p^{(k)}}.$$

Proof. As for the steepest descent method, we write that $(r^{(k+1)})^T r^{(k)} = 0$ thus we get

$$\alpha_k = \frac{(r^{(k)})^T p^{(k)}}{(p^{(k)})^T A p^{(k)}}.$$

We must check that this scalar satisfies all the orthogonality conditions. We prove this by recurrence. Indeed, $(r^{(j)})^T r^{(k+1)} = (r^{(j)})^T r^{(k)} -$

$\alpha_k (r^{(j)})^T A p^{(k)}, j \leq k - 1$ thus $(r^{(j)})^T r^{(k+1)} = 0$ because $(r^{(j)})^T r^{(k)} = 0$ and $(r^{(j)})^T A p^{(k)} = 0$.
The descent direction satisfies

$$p^{(k+1)} = r^{(k+1)} + \beta_k p^{(k)}, k \geq 0.$$

By using that $(p^{(k)})^T A p^{(k+1)} = 0$, we get

$$\beta_k = -\frac{(r^{(k+1)})^T A p^{(k)}}{(p^{(k)})^T A p^{(k)}}.$$

We now prove by recurrence that this scalar satisfies all the orthogonality properties. For $j \leq k-1$, we get $(p^{(j)})^T A p^{(k+1)} = (p^{(j)})^T A (r^{(k+1)} + \beta_k p^{(k)}) = (p^{(j)})^T A r^{(k+1)}$. Using the symmetry of A, it can be written $(p^{(j)})^T A p^{(k+1)} = (r^{(k+1)})^T A p^{(j)}$. Now, $r^{(j+1)} = r^{(j)} - \alpha_j A p^{(j)}$ thus $(p^{(j)})^T A p^{(k+1)} = 0$ by using the orthogonality of the residuals and the property $\alpha_j \neq 0$. ∎

6.4.2 CG Algorithm

In order to reduce the number of scalar products, we can define the scalars α_k and β_k in a different way. This gives finally the following definition of the Conjugate Gradient method.

Proposition 6.15 *As long as* $r^{(k)} \neq 0$*, the Conjugate Gradient iterative method defines the sequence* $\{x^{(k)}\}, k \geq 0$ *by*

$$x^{(0)} \quad given, \tag{6.34}$$
$$r^{(0)} = b - Ax^{(0)}, \tag{6.35}$$
$$p^{(0)} = r^{(0)}, \tag{6.36}$$
$$\alpha_k = \frac{(r^{(k)})^T r^{(k)}}{(p^{(k)})^T A p^{(k)}}, \tag{6.37}$$
$$x^{(k+1)} = x^{(k)} + \alpha_k p^{(k)}, \tag{6.38}$$
$$r^{(k+1)} = r^{(k)} - \alpha_k A p^{(k)}, \tag{6.39}$$
$$\beta_k = \frac{(r^{(k+1)})^T r^{(k+1)}}{(r^{(k)})^T r^{(k)}}, \tag{6.40}$$
$$p^{(k+1)} = r^{(k+1)} + \beta_k p^{(k)}. \tag{6.41}$$

It is such that $\phi(x^{(k)}) = \min_{y \in x^{(0)} + \mathcal{K}_k(A, r^{(0)})} \phi(y)$.

Proof. It remains to prove the new definition of the scalars. First, since $p^{(k)} = r^{(k)} + \beta_{k-1} p^{(k-1)}, k \geq 1$ and $p^{(0)} = r^{(0)}$, we get $(r^{(k)})^T p^{(k)} = (r^{(k)})^T r^{(k)}$ and

$$\alpha_k = \frac{(r^{(k)})^T r^{(k)}}{(p^{(k)})^T A p^{(k)}}.$$

Also, since $Ap^{(k)} = \frac{1}{\alpha_k}(r^{(k)} - r^{(k+1)})$, we get

$$\beta_k = \frac{(r^{(k+1)})^T A r^{(k+1)}}{(r^{(k)})^T r^{(k)}}.$$

The procedure for the Conjugate Gradient method is depicted in Algorithm 6.4.

Algorithm 6.4 Conjugate Gradient Algorithm

```
% x0,b given; eps1 given tolerance
% kmax  maximum number of allowed iterations,
r=b-A*x0;p=r;x=x0;
k=0;nb=norm(b);nr=r'*r;
e=sqrt(nr)/nb;
while e>eps1 & k<=kmax
    k=k+1;
    q=A*p;nap=p'*q;
    alpha=nr/nap;
    x=x+alpha*p;
    r=r-alpha*q;
    nr1=r'*r;
    beta=nr1/nr;
    p=r+beta*p;
  nr=nr1;
    e=sqrt(nr)/nb;
end
```

6.4.3 Convergence of CG

As for the steepest descent method, it is easy to prove a monotonous convergence by using the orthogonality of two consecutive residuals. The result is the same as (6.24).

But because all the residuals are here orthogonal, it is possible to prove a better convergence estimation. The proof relies on the polynomial characterization 6.2 and on the following min-max property.

Lemma 6.3 *CG iterations satisfy*

$$\|e^{(k)}\|_A = \min_{q \in \mathcal{P}_k^0} \|q(A)e^{(0)}\|_A,$$

where $\mathcal{P}_k^0$ is the set of polynomials q of degree k such that $q(0) = 1$.

Proof. This is simply a polynomial translation of the minimization property in the Krylov subspace. ∎

Proposition 6.16 *The errors satisfy*

$$\|e^{(k)}\|_A \leq \|e^{(0)}\|_A \min_{q \in \mathcal{P}_k^0} \max_{i=1,\ldots,n} |q(\lambda_i)|, \tag{6.42}$$

where λ_i, $i = 1, \ldots, n$ are the eigenvalues of A.

Proof. Let $A = U^T D U$ the spectral decomposition of A, where D is the diagonal matrix $diag(\lambda_1, \ldots, \lambda_n)$, and U is the orthogonal matrix of eigenvectors with $U^T = U^{-1}$.
Let $e^{(0)} = Uy$ then

$$\|e^{(0)}\|_A^2 = (e^{(0)})^T A e^{(0)} = y^T D y = \sum_{i=1}^n \lambda_i y_i^2.$$

Similarly,

$$\|q(A)e^{(0)}\|_A^2 = y^T q(D) D q(D) y = \sum_{i=1}^n q(\lambda_i)^2 \lambda_i y_i^2.$$

Thus

$$\|q(A)e^{(0)}\|_A^2 \leq \max_i |q(\lambda_i)|^2 \|e^{(0)}\|_A^2.$$

∎

By using Chebyshev polynomials of first kind, it is possible to prove the following result. This bound is better than the bound in (6.24).

Corollary 6.1 *CG iterations satisfy*

$$\|e^{(k)}\|_A \leq 2\|e^{(0)}\|_A \left(\frac{\sqrt{\kappa(A)} - 1}{\sqrt{\kappa(A)} + 1}\right)^k. \tag{6.43}$$

6.4.4 Preconditioned Conjugate Gradient

To improve convergence, the method must be preconditioned in practice, like the steepest descent. Again, we precondition the matrix A in writing an equivalent system to (6.1):

$$CAC^T (C^{-T} x) = Cb,$$

where C is invertible, so that CAC^T is also spd. The choice of C should be such that $\kappa(CAC^T) << \kappa(A)$. The original Conjugate Gradient Algorithm 6.4 can then be rewritten using the matrix M defined by $M^{-1} = C^T C$. This preconditioning matrix M is spd.
By introducing the preconditioned residual vector $z = M^{-1}r$, we get the preconditioned conjugate gradient (PCG) algorithm.

Proposition 6.17 *As long as $r^{(k)} \neq 0$, the Preconditioned Conjugate Gradient iterative method defines the sequence $\{x^{(k)}\}, k \geq 0$ by*

$$x^{(0)} \qquad given, \tag{6.44}$$

$$r^{(0)} = b - Ax^{(0)}, \tag{6.45}$$

$$z^{(0)} = M^{-1}r^{(0)}, \tag{6.46}$$

$$p^{(0)} = z^{(0)}, \tag{6.47}$$

$$\alpha_k = \frac{(r^{(k)})^T z^{(k)}}{(p^{(k)})^T A p^{(k)}}, \tag{6.48}$$

$$x^{(k+1)} = x^{(k)} + \alpha_k p^{(k)}, \tag{6.49}$$

$$r^{(k+1)} = r^{(k)} - \alpha_k A p^{(k)}, \tag{6.50}$$

$$z^{(k+1)} = M^{-1}r^{(k+1)}, \tag{6.51}$$

$$\beta_k = \frac{(r^{(k+1)})^T z^{(k+1)}}{(r^{(k)})^T z^{(k)}}, \tag{6.52}$$

$$p^{(k+1)} = z^{(k+1)} + \beta_k p^{(k)}. \tag{6.53}$$

6.4.5 Memory and CPU Requirements in PCG

Each iteration requires a fixed number of vector operations. The most expensive operations are the matrix-vector product and solving the preconditioning system. The total number of operations is linear with the number of iterations.

6.4.6 Relation with the Lanczos Method

CG is related to the Lanczos method defined in Chapter 5.
By normalizing the residuals, we get an orthonormal basis of the Krylov subspace, which satisfies the Lanczos relation.

Proposition 6.18 *Let $v_{j+1} = \frac{r_j}{\|r_j\|}$, $j = 0, 1, \ldots$ and $V_k = (v_1, \ldots, v_k)$. The system V_k is an orthonormal basis of the Krylov subspace $\mathcal{K}_k(A, r_0)$ which verifies the Lanczos relation*

$$AV_k = V_k T_k + \delta_k v_{k+1} e_k^T \tag{6.54}$$

where $T_k \in \mathbb{R}^{k \times k}$ is the tridiagonal matrix given by

$$T_k = \begin{pmatrix} \gamma_1 & \delta_1 & & \\ \delta_1 & \gamma_2 & \delta_2 & \\ & \cdot & \cdot & \cdot \\ & & \delta_{k-1} & \gamma_k \end{pmatrix},$$

and $e_k^T = (0 \ldots 0\ 1) \in \mathbb{R}^k$; the scalars are defined by

$$\delta_k = -\frac{\sqrt{\beta_{k-1}}}{\alpha_{k-1}}, \quad \gamma_{k+1} = \frac{1}{\alpha_k} + \frac{\beta_{k-1}}{\alpha_{k-1}}.$$

The matrix T_k is spd.

Proof. See for example reference [27]. ∎

Moreover, the iterates in CG are defined by solving a tridiagonal system.

Corollary 6.2 *Each iteration of the GC method is equivalent to compute $x_k = x_0 + V_k y$ where $y \in \mathbb{R}^k$ is solution of*

$$T_k y = \|r_0\| e_1, \tag{6.55}$$

where $u_1^T = (1\ 0 \ldots 0) \in \mathbb{R}^k$.

Proof. The subspace condition is equivalent to $x^{(k)} = x^{(0)} + V_k y$ and the orthogonality condition to $V_k^T r^{(k)} = 0$. Since $r^{(k)} = r^{(0)} - AV_k y$, the orthogonality condition can be rewritten thanks to the Lanczos relation: $T_k y = \|r_0\| u_1$.

6.4.7 Case of Symmetric Indefinite Systems: SYMMLQ Method

When A is symmetric indefinite, a breakdown can occur in CG, since it might happen that $p^{(k)} A p^{(k)} = 0$. Thus other methods are required.

When A is spd, the tridiagonal matrix T_k is also spd, non-singular, and can be factorized with a Cholesky decomposition. This is done in CG algorithm. When A is not definite, the matrix T_k can become singular, corresponding in fact to a breakdown in CG. The method SYMMLQ computes the basis V_k and the matrix T_k with the Lanczos procedure and it factorizes the matrix T_k with a LQ decomposition [49]. This can be done with recurrence relations. Breakdowns are in most cases avoided by using look-ahead techniques [12].

6.5 The Generalized Minimal Residual Method

Let A be a general matrix, not necessarily symmetric.

The generalized minimal residual method (GMRES) is a Krylov method which uses the second formulation of the subspace condition (6.12) and the general minimization property of the residual [59]. The method minimizes the Euclidian norm of the residual in the Krylov subspace, more precisely

$$\|r^{(k)}\| = \min_{z \in x^{(0)} + \mathcal{K}_k(A, r^{(0)})} \|b - Az\|. \tag{6.56}$$

The method is well defined as long as the Krylov subspace is of full dimension.

Proposition 6.19 *If $\dim(\mathcal{K}_k(A, r^{(0)})) = k$, then Problem (6.56) has a unique solution and is equivalent to the orthogonality condition*

$$(r^{(k)})^T Ay = 0, \ \forall y \in \mathcal{K}_k(A, r^{(0)})$$

Proof. Let V_k be a basis of the Krylov subspace and $z \in x^{(0)} + \mathcal{K}_k(A, r^{(0)})$. Then $b - Az = r^{(0)} - AV_k y$ with $y \in \mathbb{R}^k$ and Problem (6.56) is equivalent to the minimization problem

$$\|r^{(0)} - AV_k y_k\| = \min_{y \in \mathbb{R}^k} \|r^{(0)} - AV_k y\|.$$

Since A is non-singular and V_k is of full rank, AV_k is of full rank and the problem has a unique solution. This least-squares problem is equivalent to the orthogonality condition

$$(r^{(0)} - AV_k y_k) \perp AV_k y, \forall y \in \mathbb{R}^k,$$

or equivalently $(r^{(k)})^T A z = 0, \forall z \in \mathcal{K}_k(A, r^{(0)})$. ∎
The solution is found when the Krylov series becomes stationary.

Proposition 6.20 *If* $\dim(\mathcal{K}_k(A, r^{(0)})) = k$ *and* $\mathcal{K}_{k+1}(A, r^{(0)}) = \mathcal{K}_k(A, r^{(0)})$, *then* $r^{(k)} = 0$.

Proof. By assumption, $A^{-1}b \in x^{(0)} + \mathcal{K}_k(A, r^{(0)})$ thus, by uniqueness, it is the solution $x^{(k)}$ of problem (6.56) and $r^{(k)} = 0$. ∎

6.5.1   Krylov Basis Computation

The subspace condition in GMRES requires to compute an orthonormal Krylov basis V_k. This is achieved with the Arnoldi's procedure described in Chapter 5. Assuming that $\dim \mathcal{K}_{k+1}(A, r^{(0)}) = k + 1$, the procedure gives the Arnoldi's relation

$$AV_k = V_{k+1} \tilde{H}_k, \tag{6.57}$$

where $\tilde{H}_k \in \mathbb{R}^{(k+1) \times k}$ is an upper Hessenberg matrix. Recall that $v_1 = \frac{r^{(0)}}{\|r^{(0)}\|}$. Let $\beta = \|r^{(0)}\|$.

6.5.2   GMRES Algorithm

Now, it is possible to express the minimization property thanks to the Arnoldi's relation. Let $x = x^{(0)} + V_k y \in x^{(0)} + \mathcal{K}_k(A, r^{(0)})$, then

$$
\begin{aligned}
b - Ax &= r^{(0)} - AV_k y, \\
&= \beta v_1 - V_{k+1} \tilde{H}_k y, \\
&= V_{k+1}(\beta e_1 - \tilde{H}_k y),
\end{aligned}
$$

where e_1 is the first canonical vector of $\mathbb{R}^{k+1}$. Finally

$$\|b - Ax\| = \|\beta e_1 - \tilde{H}_k y\|$$

since V_{k+1} is an orthonormal system.

Thus minimizing the residual is equivalent to minimize $\|\beta e_1 - \tilde{H}_k y\|$ with $y \in \mathbb{R}^k$. This can be achieved by using Givens rotations, as explained in Chapter 4. The GMRES algorithm can now be fully described.

Proposition 6.21 *As long as* $\dim \mathcal{K}_{k+1}(A, r^{(0)}) = k + 1$, *the GMRES algorithm defines the sequence* $x^{(k)}$ *by*

$$x^{(k)} = x^{(0)} + AV_k y_k,$$

where V_k *is the orthonormal basis of the Krylov subspace* $\mathcal{K}_k(A, r^{(0)})$ *built by the Arnoldi's procedure. The vector* y_k *is the solution of the minimization problem*

$$\min_{y \in \mathbb{R}^k} \|\beta e_1 - \tilde{H}_k y\|,$$

where e_1 *is the first canonical vector of* $\mathbb{R}^{k+1}$ *and the Hessenberg matrix* $\tilde{H}_k$ *is defined by the Arnodi's relation*

$$AV_k = V_{k+1} \tilde{H}_k.$$

Algorithm 6.5 gives the general structure of the method.

Algorithm 6.5 GMRES Algorithm

```
% x0,b given; eps1 given tolerance
% kmax  maximum number of iterations
r0=b-A*x0;beta=norm(r0);v1=r0/beta;
k=0;nb=norm(b);e=beta/nb;
while e>eps1 & k<=kmax
   k=k+1;
   Vk=[v1 ... vk];
 %Arnoldi process
 compute vkp1 orthogonal to Vk;
   compute the last column of the matrix Htildek;
 %Arnoldi relation : AVk=Vkp1 Htildek
 %Least-squares solving
   compute yk minimizing norm(beta e1 - Htildek*y);
   nr = norm(beta e1 - Htildek*yk);
   e=nr/nb;
end
x=x0+Vk*yk;
```

6.5.3   Convergence of GMRES

The minimization property implies a monotonous convergence.

Proposition 6.22 *The residuals satisfy*

$$\|r^{(k+1)}\| \leq \|r^{(k)}\|.$$

Proof. It is evident since $x^{(k)} \in \mathcal{K}_{k+1}(A, r^{(0)})$. ∎

However, convergence is not strictly monotonous and stagnation can occur.

Proposition 6.23 *Let us assume that* $\dim(K_{k+1}) = k + 1$ *and* $r^{(k)} \neq r^{(k-1)}$. *Then* $r^{(k+1)} = r^{(k)}$ *if and only if* $(r^{(k)})^T * Ar^{(k)} = 0$.

Proof. If $r^{(k+1)} = r^{(k)}$, then $r^{(k)}$ is the unique solution of problem (6.56) for index $k + 1$ thus $r^{(k)}$ satisfies the orthogonality condition in $\mathcal{K}_{k+1}(A, r^{(0)})$. Since $r^{(k)} \in \mathcal{K}_{k+1}(A, r^{(0)})$, it gives the orthogonality property $(r^{(k)})^T Ar^{(k)} = 0$.

Converserly, assume that $(r^{(k)})^T Ar^{(k)} = 0$. If $r^{(k)} \in \mathcal{K}_k(A, r^{(0)})$, then $r^{(k)}$ is solution of problem (6.56) for index $k - 1$ and by uniqueness, $r^{(k)} = r^{(k-1)}$. Thus, $r^{(k)} \notin \mathcal{K}_k(A, r^{(0)})$ and $(r^{(k)})^T * Ay = 0, \forall y \in \mathcal{K}_{k+1}(A, r^{(0)})$. By uniqueness, $r^{(k)} = r^{(k+1)}$. ∎

Convergence can also be related to polynomials, as for CG. However, the result is not so nice.

Proposition 6.24 *If A is diagonalizable, let $A = UDU^{-1}$, where the columns of U are a basis of eigenvectors and where $D = diag(\lambda_1, \ldots, \lambda_n)$; let $\kappa(U) = \|U\|_2\|U^{-1}\|_2$ the condition number of U.*
GMRES iterations satisfy:

$$\|r^{(k)}\| \leq \|r^{(0)}\| \ \kappa(U) \ \min_{q \in \mathcal{P}_k^0} \max_{1 \leq i \leq n} |q(\lambda_i)|. \tag{6.58}$$

Proof. We have $r_k = q_k(A)r_0$; by using the diagonalization, we get $q_k(A) = Uq_k(D)U^{-1}$.
Thus $\|q_k(A)r_0\| \leq \|U\|\|q_k(D)\|\|U^{-1}\|\|r_0\|$.
Since $\|q_k(D)\| = \max_i |q_k(\lambda_i)|$, it gives inequality (6.58). ∎

6.5.4 Preconditioned GMRES

To improve convergence, the method must be preconditioned in practice, like the conjugate gradient. Here, we precondition on the right the matrix A in writing an equivalent system to (6.1):

$$AM^{-1}(Mx) = b,$$

where M is invertible. This provides the true residual in GMRES computations, whereas a left preconditioner provides a preconditioned residual. The choice of M is not easy, see Section 6.7. The GMRES algorithm is rewritten simply by replacing A with AM^{-1}.

6.5.5 Restarted GMRES

GMRES algorithm has long recurrences, in contrary to PCG. Arnoldi's procedure requires storing the $k + 1$ vectors of the basis V_{k+1}, thus the memory requirements increase at each iteration of the GMRES algorithm, whereas they are fixed in PCG.

Arnoldi's procedure requires k matrix-vector operations, thus one matrix-vector per iteration as in PCG. But the orthogonalization involves $O(nk^2)$ floating-point operations, thus the complexity is quadratic with the number k of iterations.

For very large systems arising in real scientific problems, these requirements are an issue. Thus restarted GMRES was introduced. It performs m iterations of the classical GMRES, where m is a fixed integer parameter, and it computes the approximate solution $x^{(m)}$. If $x^{(m)}$ is not accurate enough, the algorithm is restarted by taking $x^{(m)}$ as the new initial guess $x^{(0)}$.

This restarted algorithm has fixed memory requirements with $O(m)$ vectors to store. It has also a complexity linear with the number of restarts.

The restarting step m must be chosen wisely for each problem, otherwise the restarted GMRES(m) might stagnate. In order to avoid stagnation, deflation techniques and augmented techniques have been designed [48].

6.5.6 MINRES Algorithm

When the matrix A is symmetric, the Arnoldi relation becomes the Lanczos relation and the Hessenberg matrix is a symmetric tridiagonal matrix. If A is indefinite, breakdowns could occur in PCG. The GMRES algorithm can then be simplified and short recurrences can be used to solve the small least squares problem. This algorithm for symmetric indefinite matrices is called MINRES.

6.6 The Bi-Conjugate Gradient Method

The Bi-Conjugate Gradient method (BiCG) is a Krylov projection method. It is defined by the subspace condition (6.10) and by the orthogonality condition:

$$(r^{(k)})^T v = 0, \ \forall v \in \mathcal{K}_k(A^T, \tilde{r}^{(0)}) \tag{6.59}$$

where $\tilde{r}^{(0)}$ is a given vector. In order to satisfy this condition, the algorithm also solves a shadow linear system with the matrix transpose. The algorithm was first introduced by Lanczos in 1952 [43] and reformulated as a Conjugate Gradient-like method by Fletcher in 1974 [29].

So the subspace conditions are

$$\begin{aligned} x^{(k+1)} &= x^{(k)} + \alpha_k p^{(k)}, \\ \tilde{x}^{(k+1)} &= \tilde{x}^{(k)} + \tilde{\alpha}_k \tilde{p}^{(k)}, \end{aligned}$$

and the orthogonality conditions are

$$\begin{aligned} (r^{(k)})^T v &= 0, \ \forall v \in \mathcal{K}_k(A^T, \tilde{r}^{(0)}), \\ (\tilde{r}^{(k)})^T v &= 0, \ \forall v \in \mathcal{K}_k(A, r^{(0)}). \end{aligned}$$

In practice, and from now on, $\tilde{r}^{(0)} = r^{(0)}$. Also, $p^{(0)} = r^{(0)} = \tilde{p}^{(0)}$.
Because the matrix is not spd, breakdowns can occur during the algorithm.
A breakdown occurs when $(r^{(k)})^T \tilde{r}^{(k)} = 0$. Another breakdown occurs when $(\tilde{p}^{(k)})^T A p^{(k)} = 0$.

6.6.1   Orthogonality Properties in BiCG

As in CG, the residuals form a basis of the Krylov subpsace associated with A, and the shadow residuals form also a basis of the Krylov subspace associated with A^T. Here, the two basis are bi-orthogonal. However, it requires assuming no breakdown.

Proposition 6.25 *If $(\tilde{r}^{(j)})^T r^{(j)} \neq 0, 0 \leq j \leq k-1$, then the sequence of residuals $\{r^{(0)}, \ldots, r^{(k-1)}\}$ obtained from the Bi-Conjugate Gradient method is a basis of the Krylov subspace $\mathcal{K}_k(A, r^{(0)})$. Also, the sequence of residuals $\{\tilde{r}^{(0)}, \ldots, \tilde{r}^{(k-1)}\}$ obtained from the shadow system is a basis of the Krylov subspace $\mathcal{K}_k(A^T, r^{(0)})$. The two basis are bi-orthogonal: $(\tilde{r}^{(i)})^T r^{(j)} = 0$, $i \neq j$, $0 \leq i \leq k$, $0 \leq j \leq k$.*

Proof. Let $0 \leq j \leq k-1$ and $0 \leq i \leq j-1$; since $\tilde{r}^{(i)} \in \mathcal{K}_k(A^T, \tilde{r}^{(0)})$, orthogonality conditions imply that $(\tilde{r}^{(i)})^T r^{(j)} = 0$.
Let $i \geq j-1$; since $r^{(j)} \in \mathcal{K}_k(A, r^{(0)})$, orthogonality conditions imply that $(\tilde{r}^{(i)})^T r^{(j)} = 0$.
Thus the two systems are bi-orthogonal. It remains to prove that each system is linearly independent. Let us consider a linear combination of the residuals equal to zero: $\sum_{i=0}^{k-1} \beta_i r_i = 0$; by multiplying par $\tilde{r}^{(j)})^T$, we get $\beta_j (\tilde{r}^{(j)})^T r^{(j)} = 0$. The assumption of no breakdown implies then that $\beta_j = 0$. Thus the system of residuals forms a basis of the Krylov subspace.
A similar proof allows concluding that the system of shadow residuals form a basis. $\blacksquare$
As in CG, the descent directions form also a basis of the Krylov subpsaces, provided there is no breakdown.

Proposition 6.26 *If $(\tilde{p}^{(j)})^T A p^{(j)} \neq 0, 0 \leq j \leq k-1$, then the sequence of descent directions $\{p^{(0)}, \ldots, p^{(k-1)}\}$ obtained from the Bi-Conjugate Gradient method is a basis of the Krylov subspace $\mathcal{K}_k(A, r^{(0)})$. Also, the sequence* of descent directions $\{\tilde{p}^{(0)}, \ldots, \tilde{p}^{(k-1)}\}$ obtained from the shadow system is a basis of the Krylov subspace $\mathcal{K}_k(A^T, r^{(0)})$. The two basis are bi-orthogonal: $(\tilde{p}^{(i)})^T p^{(j)} = 0$, $i \neq j$, $0 \leq i \leq k$, $0 \leq j \leq k$.

Proof. We first remark that the subspace conditions and the orthogonality conditions imply that Ap_k is orthogonal to the shadow Krylov subspace $\mathcal{K}_k(A^T, r^{(0)})$ and that $A^T \tilde{p}_k$ is orthogonal to the Krylov subspace $\mathcal{K}_k(A, r^{(0)})$. Consequently, we get $(\tilde{p}^{(i)})^T A p^{(j)} = 0$, for $i \neq j$.

The same proof as above shows that the system of descent directions is linearly independent, thanks to the assumption of no breakdown. For the same reason, the system of shadow descent directions is linearly independent. Thus they each form a basis. $\blacksquare$

Now, by using the basis properties, it is possible to find a short recurrence relation to compute the descent directions; the process follows the same lines as in CG.

Proposition 6.27 *Assuming no breakdown, the descent directions are given by:*
$$p^{(k)} = r^{(k)} + \beta_{k-1} p^{(k-1)}, \quad \tilde{p}^{(k)} = \tilde{r}^{(k)} + \tilde{\beta}_{k-1} \tilde{p}^{(k-1)}.$$

Proof. The proof follows the same lines as for CG.

Since the residuals form a basis, $r^{(k)} \notin \mathcal{K}_k(A, r^{(0)})$ and the system $\{p^{(0)}, \ldots, p^{(k-1)}, r^{(k)}\}$ is a basis of $\mathcal{K}_{k+1}(A, r^{(0)})$.

Thus $p^{(k)} = \sum_{i=0}^{k-1} \beta_i p^{(i)} + \beta_k r^{(k)}$.

Since the descent directions form a basis, $p^{(k)} \notin \mathcal{K}_k(A, r^{(0)})$ thus $\beta_k \neq 0$; we can choose $\beta_k = 1$.

Let $0 \leq j \leq k-2$; we have $(\tilde{p}^{(j)})^T A p^{(k)} = 0$ and $(\tilde{p}^{(j)})^T A p^{(i)} = 0, i \neq j$ and $(\tilde{p}^{(j)})^T A r^{(k)} = 0$ and $(\tilde{p}^{(j)})^T A p^{(j)} \neq 0$ thus $\beta_j = 0$.

The proof is similar for the shadow system. $\blacksquare$

Orthogonality conditions give the values of the scalar α_k and β_k.

Proposition 6.28 *Assuming no breakdown, the scalars are given by:*
$$\alpha_k = \frac{(\tilde{r}^{(k)})^T r^{(k)}}{(\tilde{p}^{(k)})^T A p^{(k)}},$$
$$\tilde{\alpha}_k = \alpha_k,$$
$$\beta_k = \frac{(\tilde{r}^{(k+1)})^T r^{(k+1)}}{(\tilde{r}^{(k)})^T r^{(k)}},$$
$$\tilde{\beta}_k = \beta_k.$$

Proof. The proof is the same as for CG.

By writing $r^{(k+1)} = r^{(k)} - \alpha_k A p^{(k)}$ and $(\tilde{r}^{(k)})^T r^{(k+1)} = 0$,

we get:
$$\alpha_k = \frac{(\tilde{r}^{(k)})^T r^{(k)}}{(\tilde{r}^{(k)})^T A p^{(k)}}.$$

But $\tilde{p}^{(k)} = \tilde{r}^{(k)} + \tilde{\beta}_{k-1} \tilde{p}^{(k-1)}$. Thus $(\tilde{p}^{(k)})^T A p^{(k)} = (\tilde{r}^{(k)}) A p^{(k)}$.

As for the shadow system, we get by symmetry and transposition $\tilde{\alpha}_k = \frac{(\tilde{p}^{(k)})^T A r^{(k+1)}}{(\tilde{p}^{(k)})^T A p^{(k)}}$. Thus $\tilde{\alpha}_k = \alpha_k$.

By writing $p^{(k+1)} = r^{(k+1)} + \beta_k p^{(k)}$ and $(\tilde{p}^{(k)})^T A p^{(k+1)} = 0$,

we get: $\beta_k = \frac{(\tilde{r}^{(k)})^T r^{(k)}}{(\tilde{r}^{(k)})^T A p^{(k)}}.$

But $\tilde{r}^{(k+1)} = \tilde{r}^{(k)} - \alpha_k A^T \tilde{p}^{(k)}$.

Thus $(r^{(k+1)})^T \tilde{r}^{(k+1)} = -\alpha_k (r^{(k+1)})^T A^T \tilde{p}^{(k)}$

and $\beta_k = \frac{(\tilde{r}^{(k+1)})^T r^{(k+1)}}{(\tilde{r}^{(k)})^T r^{(k)}}$.

By symmetry, $\tilde{\beta}_k = \beta_k$.

It remains to prove that all orthogonality conditions are satisfied. We proceed by recurrence, as for CG algorithm. ∎

6.6.2   BiCG Algorithm

Proposition 6.29 *As long as there is no breakdown, the Bi-Conjugate Gradient iterative method defines the sequence* $\{x^{(k)}\}, k \geq 0$ *by:*

$$x^{(0)} \text{ given,}$$
$$r^{(0)} = b - Ax^{(0)}, \tilde{r}^{(0)} = r^{(0)},$$
$$p^{(0)} = r^{(0)}, \tilde{p}^{(0)} = r^{(0)},$$
$$\alpha_k = \frac{(\tilde{r}^{(k)})^T r^{(k)}}{(\tilde{p}^{(k)})^T A p^{(k)}},$$
$$x^{(k+1)} = x^{(k)} + \alpha_k p^{(k)},$$
$$r^{(k+1)} = r^{(k)} - \alpha_k A p^{(k)}, \tilde{r}^{(k+1)} = \tilde{r}^{(k)} - \alpha_k A^T \tilde{p}^{(k)},$$
$$\beta_k = \frac{(\tilde{r}^{(k+1)})^T r^{(k+1)}}{(\tilde{r}^{(k)})^T r^{(k)}},$$
$$p^{(k+1)} = r^{(k+1)} + \beta_k p^{(k)}, \tilde{p}^{(k+1)} = \tilde{r}^{(k+1)} + \beta_k \tilde{p}^{(k)}.$$

The bi-conjugate gradient procedure is depicted in Algorithm 6.6.

Algorithm 6.6 Bi-Conjugate Gradient Algorithm

```
% x0,b given; eps1 given tolerance
% kmax  maximum number of iterations,
r=b-A*x0;beta=0;r1=r;p=0;p1=0;
x=x0;k=0;e=1;nb=norm(b);
while e>eps1 & k<=kmax
   k=k+1;p=r+beta*p;p1=r1+beta*p1;
   q=A*p;q1=A^T p1;
   nr=r1'*r;nap=p1'*q;alpha=nr/nap;
   x=x+alpha*p;r=r-alpha*q;
   r1=r1-alpha*q1;nr1=r1'*r;beta=nr1/nr;
   nr=nr1;e=norm(r)/nb;
end
```

6.6.3   Convergence of BiCG

Contrary to GMRES, the BiCG algorithm provides a short recurrence but is not related to a minimization property. Therefore the convergence is not monotonous. In practice, it exhibits an erratic behavior. This irregular, erratic behavior may slow down the speed of convergence.

6.6.4 Breakdowns and Near-Breakdowns in BiCG

Building the algorithm assumes no breakdown; as can be seen in the algorithm description, the breakdowns induce a division by zero in the iteration. When the divisors are close to zero, a near-breakdown occurs and instability may be expected. These situations are studied for example in references [30] and [54], where look-ahead techniques are proposed.

6.6.5 Complexity of BiCG and Variants of BiCG

Each iteration requires two matrix-vector products, one by A and one by A^T. Thus each iteration is about twice as expensive as a CG iteration. Clearly, CG should be preferred when the matrix is spd.

In some cases, the product by A^T is not as fast as the product by A. Therefore, transpose free variants were designed, such as the Conjugate Gradient Squared (CGS) algorithm [63].

However, CGS still suffers from an erratic convergence. The stabilized algorithm BiCGSTAB was designed to provide a smoother convergence [70]. Similar algorithms, called BiCGSTAB(l), were also investigated [62].

6.6.6 Preconditioned BiCG

In practice, the BiCG algorithm uses a preconditioned version. Let M be a non-singular matrix; the preconditioned algorithm involves two linear systems, with M and with M^T.

Proposition 6.30 *As long as there is no breakdown, the preconditioned Bi-Conjugate Gradient iterative method defines the sequence* $\{x^{(k)}\}, k \geq 0$ *by:*

$$x^{(0)} \ given,$$
$$r^{(0)} = b - Ax^{(0)}, \tilde{r}^{(0)} = r^{(0)},$$
$$z^{(0)} = M^{-1}r^{(0)}, \tilde{z}^{(0)} = M^{-T}\tilde{r}^{(0)},$$
$$p^{(0)} = z^{(0)}, \tilde{p}^{(0)} = \tilde{z}^{(0)},$$
$$\alpha_k = \frac{(\tilde{r}^{(k)})^T z^{(k)}}{(\tilde{p}^{(k)})^T A p^{(k)}},$$
$$x^{(k+1)} = x^{(k)} + \alpha_k p^{(k)},$$
$$r^{(k+1)} = r^{(k)} - \alpha_k A p^{(k)}, \tilde{r}^{(k+1)} = \tilde{r}^{(k)} - \alpha_k A^T \tilde{p}^{(k)},$$
$$z^{(k+1)} = M^{-1}r^{(k+1)}, \tilde{z}^{(k+1)} = M^{-T}\tilde{r}^{(k+1)},$$
$$\beta_k = \frac{(\tilde{r}^{(k+1)})^T z^{(k+1)}}{(\tilde{r}^{(k)})^T z^{(k)}},$$
$$p^{(k+1)} = z^{(k+1)} + \beta_k p^{(k)}, \tilde{p}^{(k+1)} = \tilde{z}^{(k+1)} + \beta_k \tilde{p}^{(k)}.$$

The procedure is depicted in Algorithm 6.7.

Algorithm 6.7 Preconditioned Bi-Conjugate Gradient Algorithm

```
% x0,b given; eps1 given tolerance
% kmax   maximum number of iterations,
r=b-A*x0;beta=0;r1=r;p=0;p1=0;
x=x0;k=0;e=1;nb=norm(b);
while e>eps1 & k<=kmax
    k=k+1;z=M\r;p=z+beta*p;
    z1=M^T \ r1;p1=z1+beta*p1;
    q=A*p;q1=A^T p1;
    nr=r1'*z;nap=p1'*q;alpha=nr/nap;
    x=x+alpha*p;r=r-alpha*q;
    r1=r1-alpha*q1;nr1=r1'*z;
    beta=nr1/nr;nr=nr1;e=norm(r)/nb;
end
```

6.7   Preconditioning Issues

Preconditioning is essential in practice in order to get a competitive Krylov method with fast convergence [9]. A first class of preconditioning is based on the splitting techniques of stationary iterative methods, like Jacobi or SSOR preconditioning. However, they are not always efficient for very large systems. A second class of preconditioning relies on an incomplete factorization; this has the advantage of avoiding the fill-in occuring in complete factorizations. Again, the efficiency decreases in general when the size of the system becomes large. Approximate inverse techniques suffer from the same drawback.

Nowadays, two-level preconditioning approaches are the most promising methods for very large systems [67] and [68]. Multigrid is very efficient for some large systems. Coupled with PCG, it is more robust than multigrid by itself and it speeds-up convergence of the Conjugate Gradient. Domain decomposition is very often used, combined with a kind of coarse grid correction. For example, restricted Schwarz preconditioning is often used with GMRES or BiCGSTAB.

6.8 Exercises

Exercise 6.1

Analyze the complexity of SOR Algorithm 6.1 in terms of the number of operations and the memory requirements.

Exercise 6.2

Let $x = A^{-1}b$. Prove the inequality:

$$\frac{\|x - y\|}{\|x\|} \leq \kappa(A) \frac{\|b - Ay\|}{\|b\|}.$$

In view of this inequality, explain the convergence criteria used in the algorithms of this chapter.

Exercise 6.3

Prove Proposition 6.17. Write the corresponding PCG algorithm.

Exercise 6.4

Write the restarted GMRES(m) algorithm. Analyze the complexity in terms of number of operations and memory requirements.

Exercise 6.5

The objective of this exercise is to prove inequality (6.43).

1. Show that, for any $q \in \mathcal{P}_k^0$,

$$\|e^{(k)}\|_A \leq \|e^{(0)}\|_A \max_{\lambda_1 \leq t \leq \lambda_n} |q(t)|.$$

2. The idea is to choose particular polynomials q_k for which the bound can be computed. These polynomials are based on the Chebyshev polynomials of first kind, which are defined in the following way:

$$\begin{cases} T_{k+1}(X) + T_{k-1}(X) = 2XT_k(X), \\ T_0(X) = 1, T_1(X) = X. \end{cases}$$

We start by studying $T_k(\xi)$ for $\xi \in [-1, 1]$. Let us define

$$F_k(\xi) = \cos(k \arccos(\xi)), \ \forall \xi \in [-1, 1].$$

Show that $|F_k(\xi)| \leq 1$, $\forall \xi \in [-1, 1]$.

3. Show that $F_0(\xi) = 1$, $F_1(\xi) = \xi$, $\forall \xi \in [-1, 1]$.

4. Show that $F_{k+1}(\xi) + F_{k-1}(\xi) = 2\xi F_k(\xi)$, $\forall \xi \in [-1,1]$.

5. Show that $T_k(\xi) = F_k(\xi)$, $\forall \xi \in [-1,1]$.

6. We now study $T_k(\xi)$ for $\xi \geq 1$. Show that

$$T_k(\xi) = \frac{1}{2}((\xi + \sqrt{\xi^2 - 1})^k + (\xi + \sqrt{\xi^2 - 1})^{-k}), \ \forall \xi \geq 1.$$

   Indication: consider the characteristic equation $x^2 - 2\xi x + 1 = 0$.

7. Deduce that $T_k(\xi) \geq \frac{1}{2}(\xi + \sqrt{\xi^2 - 1})^k$, $\forall \xi \geq 1$.

8. We are now able to choose the polynomials q_k, defined by

$$q_k(X) = \frac{1}{T_k(\frac{\lambda_n + \lambda_1}{\lambda_n - \lambda_1})} T_k(\frac{\lambda_n + \lambda_1 - 2X}{\lambda_n - \lambda_1}).$$

   Show that $q_k(0) = 1$ and that $q_k \in \mathcal{P}_k^0$.

9. The objective is to bound $q_k(t)$, $\forall t \in [\lambda_1, \lambda_n]$. We define the following change of variable:

$$\xi = \frac{\lambda_n + \lambda_1 - 2t}{\lambda_n - \lambda_1}$$

   Show that $\xi \in [-1,1]$, $\forall t \in [\lambda_1, \lambda_n]$.

10. Let $\mu = \frac{\lambda_n + \lambda_1}{\lambda_n - \lambda_1}$. Show that $\max_{\lambda_1 \leq t \leq \lambda_n} |q_k(t)| \leq \frac{1}{T_k(\mu)}$.

11. Show that $\mu > 1$. Deduce that

$$T_k(\mu) \geq \frac{1}{2}(\frac{\sqrt{\kappa}+1}{\sqrt{\kappa}-1})^k.$$

12. Deduce that

$$\|e^{(k)}\|_A \leq 2\|e^{(0)}\|_A (\frac{\sqrt{\kappa}-1}{\sqrt{\kappa}+1})^k.$$

Chapter 7

Sparse Systems to Solve Poisson Differential Equations

In this chapter we consider sparse systems of linear equations resulting from the discretization by finite difference and finite element methods of ordinary and partial differential equations. Typically, these discretization methods are illustrated on one-dimensional and two-dimensional Poisson type differential equations. There are multiple references to these methods, of which we only cite [4], [17], [34], [39], [66] and [72].

7.1 Poisson Differential Equations

The starting point is a domain $\Omega \subset \mathbb{R}^d$, $d = 1, 2$. For the one-dimensional case $(d = 1)$, $\Omega = (0, L)$, $L > 0$ and is assumed to be an open-bounded subset of class C^1 (or piecewise C^1), with boundary $\Gamma = \partial\Omega$ and in 2 dimensions, Ω is assumed to have a "smooth" boundary. This can be characterized as follows.

Definition 7.1 *Let Ω be an open-bounded domain of $\mathbb{R}^2$. Its boundary $\Gamma = \partial\Omega$ is considered to be smooth if it can be described by a finite sequence of vertices $\{P_i | i = 1, ..., k\}$, where $(P_1 = P_k)$ and correspondingly a sequence of simple arcs $\{\overrightarrow{P_i P_{i+1}} | i = 1, 2, ..., k - 1\}$, whereby each arc $\overrightarrow{P_i P_{i+1}}$ is being described by a regular at least C^1 function $\psi_{i,i+1}(x, y) = 0$.*

Usually, ordinary and partial differential equations are given in classical forms. Solutions are assumed to belong to continuously differentiable sets of functions, $C^k(\Omega)$, $k \geq 2$.

1. **One-dimensional case**:

   Let a and b be two real-valued functions on the interval $[0, L]$, with $a(x) \geq a_0 > 0$ and $b(x) \geq 0$. Consider the one-dimensional boundary value problem:

   Find $u : \overline{\Omega} \longrightarrow \mathbb{R}$, such that:

   $$\mathcal{L}[u](x) = -\frac{d}{dx}\left(a(x)\frac{du}{dx}\right) + b(x)u = f(x), \forall\, x \in \Omega, \qquad (7.1)$$

The boundary conditions in (7.1) can be of:
1. Dirichlet type:

$$u(0) = \alpha, \ u(1) = \beta. \tag{7.2}$$

Other valid boundary conditions may be as follows:
2. Neumann's boundary conditions:

$$u'(0) = \alpha \quad \text{and} \quad u'(L) = \beta, \text{ or} \tag{7.3}$$

3. Periodic boundary conditions:

$$u(0) = u(L) \quad \text{and} \quad u'(0) = u'(L), \text{ or also} \tag{7.4}$$

4. Mixed boundary conditions:

$$c_1 u'(0) + c_2 u(0) = \alpha \quad \text{and} \quad d_1 u'(L) + d_2 u(L) = \beta, \tag{7.5}$$

with $\max\{|c_1|, |c_2|\} \neq 0$ and $\max\{|d_1|, |d_2|\} \neq 0$.

2. **Two-dimensional case**:
   Consider Poisson two-dimensional partial differential equation with variable coefficients and mixed boundary conditions, in which one seeks $u : \overline{\Omega} \longrightarrow \mathbb{R}$, such that:

$$\begin{cases} \mathcal{L}[u](x,y) = -\nabla(a(x,y)\nabla u(x,y)) = f(x,y), \forall\,(x,y) \in \Omega \\ u(x,y) = g(x,y), \forall\,(x,y) \in \Gamma_D \subset \partial\Omega, \\ \nu.\nabla u(x,y) - k(x,y)u(x,y) = h(x,y), \forall\,(x,y) \in \partial\Omega - \Gamma_D. \end{cases} \tag{7.6}$$

where a and f are given functions on Ω and g, h and k are defined on $\partial\Omega$.
A particular case of (7.6), is the case when $a(x,y) = 1$ and $\Gamma_D = \Gamma$ where one handles the two-dimensional Dirichlet-Poisson equation:

$$\begin{cases} \mathcal{L}[u](x,y) = -\Delta u(x,y) = f(x,y), \forall\,(x,y) \in \Omega \\ u(x,y) = g(x,y), \forall\,(x,y) \in \Gamma = \partial\Omega. \end{cases} \tag{7.7}$$

with $\Delta u(x,y) = \frac{\partial^2 u}{\partial x^2} + \frac{\partial^2 u}{\partial y^2}$.
When $f = 0$, (7.7) reduces to Laplace-Dirichlet's equation:

$$\begin{cases} \mathcal{L}[u](x,y) = -\Delta u(x,y) = 0, \forall\,(x,y) \in \Omega \\ u(x,y) = g(x,y), \forall\,(x,y) \in \Gamma = \partial\Omega. \end{cases} \tag{7.8}$$

On the basis of (7.1), (7.7) and (7.6), a boundary value Poisson type differential equation can be put in the form:

$$\begin{cases} \mathcal{L}[u](x) = f(x,y), \forall\,(x,y) \in \Omega \\ u(x,y) = g(x,y), \forall\,(x,y) \in \Gamma_D \subset \partial\Omega, \\ \mathcal{B}[u](x,y) = 0, (x,y) \in \partial\Omega - \Gamma_D. \end{cases} \tag{7.9}$$

7.2   The Path to Poisson Solvers

Whether using finite differences (FD) or finite element (FE) methods, obtaining a software that solves a general Poisson type equation such as (7.9) requires the following steps:

1. **Discretizing the domain $\overline{\Omega}$** a step that leads to a finite set of points ("Nodes") $\overline{\Omega}_N \subset \overline{\Omega}$ that covers $\overline{\Omega}$, with $N \to \infty$ being the discretizing parameter.
   This step is of prime importance and requires writing software that creates the needed data structures for $\overline{\Omega}_N$ and allows to proceed to the next steps.

2. Writing a discrete Poisson system on $\overline{\Omega}_N$ relative to the method being used.

3. Transforming the discrete Poisson system into an algebraic system of linear equations $AU = F$, where $A \in \mathbb{R}^{N \times N}$ and $F \in \mathbb{R}^N$. The solution U of the system is expected to approximate the unknown part of the exact solution u on N nodes of $\overline{\Omega}_N$.
   This step is crucial as it requires writing software that generates both the sparse matrix A and the vector F.

4. Solving the system $AU = F$ by use of either a direct method (as in Chapter 3) or by an iterative method (as in Chapter 6).

7.3   Finite Differences for Poisson-Dirichlet Problems

Finite difference approaches are simple to formulate and to implement, particularly in one-dimensional problems and in two-dimensional ones when the domain Ω is rectangular or a union of rectangular subdomains and the boundary conditions are of the Dirichlet type, i.e., $\Gamma_D = \Gamma$.
Thus, we assume in this section that either $\Omega = (0, L)$, $\Gamma = \{0, L\}$ (one-dimension) or

$$\Omega = (0, L) \times (0, l), \ \Gamma = [0, L] \times \{0\} \cup [0, L] \times \{l\} \cup \{0\} \times [0, l] \cup \{L\} \times [0, l].$$

A finite-difference discretization consists in replacing the derivatives in (7.1) by divided differences.

Finite Difference Formulae

Finite difference discretizations for the Poisson equations (7.1) and (7.7) are based on the central difference formulae.
Let $v \in C^k(0,a)$. For $x, x+h, x-h \in (0,a)$, one has the central difference formulae for approximating the first and second derivatives respectively:

$$\text{if } \delta_h[v(x)] = \frac{v(x+h/2) - v(x-h/2)}{h}, \tag{7.10}$$

then:

$$v'(x) = \delta_h[v(x)] + O(h^{k-1}), \; v \in C^k, 2 \le k \le 3.$$

Similarly, if

$$\delta_h^2[v(x)] = \frac{v(x+h) - 2v(x) + v(x-h)}{h^2} \tag{7.11}$$

then:

$$v''(x) = \delta_h^2[v(x)] + O(h^{k-2}), \; v \in C^k, 3 \le k \le 4.$$

More generally, for a function $a(x)$ that is sufficiently regular, one has:

$$\delta_h[a(x)\delta_h[v(x)]] =$$

$$= \frac{a(x+\frac{h}{2})(v(x+h) - v(x)) + a(x-\frac{h}{2})(v(x-h) - v(x))}{h^2}. \tag{7.12}$$

7.3.1   One-Dimensional Dirichlet-Poisson

With $Nh = L$, N integer, consider the discrete domains:

$$\Omega_h = \{x_i = ih \,|\, i = 1...N-1\}, \; \overline{\Omega}_h = \{x_i = ih \,|\, i = 0,1,...,N\}, \tag{7.13}$$

that uniformly partitioned Ω and $\overline{\Omega}$. Let $a_i = a(x_i)$ and $b_i = b(x_i)$. The discrete system corresponding to (7.1) is defined as follows:
Find $U : \overline{\Omega}_h \longrightarrow \mathbb{R}$, such that:

$$\begin{cases} \mathcal{L}_h[U](x_i) = -\delta_h[a(x_i)\delta_h[U_i]] + b_iU_i = f_i = f(x_i), \, x_i \in \Omega_h, \\ U_0 = \alpha, \, U_N = \beta. \end{cases} \tag{7.14}$$

Implementation

To construct the algebraic system that implements (7.14), we construct the discrete data structure corresponding to (7.13) and represented by Table 7.1. Specifically to each node P of the discrete domain $\overline{\Omega}_h$, we associate the following attributes:

- x_P, the abscissa of P.
- id_P that indicates whether P is a Dirichlet boundary point or an interior one. We use $id(P) = 0$ if $P \in \Gamma = \{0, L\}$ and $id(P) = 1$ if $P \in \Omega = (0, L)$.
- $a(P)$ and $f(P)$, particularly needed when $id_P = 1$.
- $U(P)$, the last attribute which is by default g when P is a boundary point $(id_P = 0)$ and otherwise, provides the values of the solution U to the discrete system when $id(P) = 1$. The MATLAB procedure 7.1 implements the construction of Table 7.1. For this one-dimensional model, note that the solution to (7.14) depends on M parameters, of which $N = M - 2$, $[U_2, ...U_{N-1}]^T$ are unknowns, since $[U_1 = \alpha \, U_M = \beta]$ are given.

I Node index	x_P	id	a	b	f	U (= g on Γ_D)
not stored						Discrete system solution
1	0	0				α
..	..	1	..	..	..	..
..	..	..	..	..	..	..
..	..	1	..	..	..	..
N	L	0				β

TABLE 7.1: One-dimensional Nodes structure.

Algorithm 7.1 Generating a One-Dimensional Nodes Structure

```
function [xp,id,a,b,f,U]= Nodes1D(pih)
% This function generates the table  Nodes
% INPUT: pih the partition of the closed interval [0,L]
% OUTPUT: The node attributes:
%         xp: Abscissa of node
%         id:  Node id (1 if interior and 0 if boundary point)
%         a,b,f: Values of a, b and f at node
%         U: Solver solution assigned to the Dirichlet data at boundary
%%%%%%%%%%%%%%%%%%%%%%%%%%%%%%%%%%%
%      Generation of the table is done through probing the grid points
%      I the node index is initialized to zero
I=0;L=max(pih);M=length(pih);
for i=1:M
         I=I+1;xp(I)=pih(i);
         if i==1 |i==M
                 id(I)=0;%Assign values of other attributes a,b, f, g
         else
                 id(I)=1;%Assign values of other attributes a, b, f
         end
end
```

Thus, the resulting system obtained from (7.14) takes the following matrix form:

$$AU = F, \tag{7.15}$$

the matrix $A \in \mathbb{R}^{N,N}$ being tri-diagonal. In case, $a(x) = 1$ and $b(x) = 0$, A is the well-known "central difference matrix":

$$A = \frac{1}{h^2} \begin{pmatrix} 2 & -1 & 0 & \cdots & 0 \\ -1 & 2 & -1 & 0 & 0 \\ \cdots & \cdots & \cdots & \cdots & \cdots \\ \cdots & 0 & -1 & 2 & -1 \\ 0 & 0 & \cdots & -1 & 2 \end{pmatrix}, \quad F = \begin{pmatrix} f_1 + \alpha/h^2 \\ f_2 \\ \cdots \\ f_{M-1} \\ f_M + \beta/h^2 \end{pmatrix}.$$

In that well-known case and other more general cases, storing the matrix A using the `sparse` commands facilities of `MATLAB`, we need to construct a data structure of its coefficients.

The `sparse` `MATLAB` environment which handles efficiently sparse matrices as discussed in Chapter 1, Section 1.7, using compressed storage by coordinates (COO). Such storage includes the following elements:

1. One starts by constructing a three-column table that stores respectively a row index i, a column index j and the non-zero coefficient $s = s_{ij}$. This is shown in Table 7.2.

i	**j**	**s**
i_1	j_1	s_1
i_2	j_2	s_2
...	...	...
i_{nnz}	j_{nnz}	s_{nnz}

TABLE 7.2: Storage of `MATLAB` sparse matrices.

2. Based on Table 7.2, the command `S = sparse(i,j,s,m,n,nzmax)` generates an m by n matrix S with $s_l = S(i_l, j_l), 1 \leq l \leq nnz, 1 \leq i_l \leq m, 1 \leq j_l \leq n$.
   Furthermore, $nnz \leq nzmax$, with $nzmax$ the maximal number of allocated memory words for the structure S.
   Note when $i_k = i_l, j_k = j_l, k \neq l$, the corresponding entries are then added.
   Furthermore, such command may have several simplifications:

   - `S = sparse(i,j,s,m,n)` uses `nzmax = length(s)`.

   - `S = sparse(i,j,s)` uses `m = max(i)` and `n = max(j)`.

   - `S = sparse(m,n)` is an abbreviation of `sparse([],[],[],m,n,0)`. It generates a zero m by n ultimate sparse matrix.

On that basis, handling the system (7.14) begins by writing it in the form:

$$-\delta_h a(x_i)\delta_h U_i + b_i U_i = \ldots$$

$$\ldots A_{P,i}Ui + A_{O,i}U_{i-1} + A_{E,i}U_{i+1} = f_i, \text{ if } \texttt{id}(i) = 1, \qquad (7.16)$$

where, by using $a(x \pm \frac{h}{2}) \approx \frac{1}{2}(a(x) + a(x \pm h))$, one has the formulae:

$$\begin{cases} A_{O,i} = -\frac{1}{2h^2}a(x_i - \frac{h}{2}) \approx -\frac{1}{2h^2}(a_i + a_{i-1}) \\ A_{E,i} = -\frac{1}{2h^2}a(x_i + \frac{h}{2}) \approx -\frac{1}{2h^2}(a_i + a_{i+1}) \\ A_{P,i} = -A_{O,i} - A_{E,i} + b_i. \end{cases} \qquad (7.17)$$

Hence, using (7.17), Algorithm 7.2 generates Table 7.3.

Algorithm 7.2 Matrix Coefficients for One-Dimensional Finite Difference

```
function [I,J,s,UNK,IB]=coeffa1(pih)
%Pt is the coeffa table pointer. It is initialized to 0
[xp,id,a,f,U]= Nodes1D(pih);
M=length(pih);Pt=0;
H=diff(pih);h=max(H);
% For a uniform partition h is the distance
%         between 2 consecutive points
%Filling the table coeffa
 for i=1:M
         if id(i)==1
                 Pt=Pt+1;I(Pt)=i;J(Pt)=i;
                 s(Pt)=(2*a(i)+a(i+1)+a(i-1))/(h*h)+b(i);
                 Pt=Pt+1;I(Pt)=i;J(Pt)=i-1;
                 s(Pt)=-(a(i)+a(i-1))/(h*h);
                 Pt=Pt+1;I(Pt)=i;J(Pt)=i+1;
                 s(Pt)=-(a(i)+a(i+1))/(h*h);
         end
 end
% UNK is the set of indices corresponding to the unknown
%    part of U
UNK=find(id==1);
% IB is the set of boundary indices such that
% there exists interior  Nodes for which A(I,J) is non-zero
IB=find(id==0);
```

Coefficient table to obtain sparse matrix for a one-dimensional FD			
Pt Pointer to non-stored index in table NODE	*I*	*J*	s_{IJ}
1			
I			
nnz			

TABLE 7.3: Matrix coefficients for one-dimensional finite difference.

At that point, Algorithm 7.3 generates the matrix A and the vector F of the system $Ax = F$ as a last stage in obtaining a finite difference solver to (7.1).

Algorithm 7.3 Sparse Matrix for One-Dimensional Finite Difference

```
function [A,F]=generate(pih)
[I,J,s,IB,UNK]=coeffa1(pih);
AG=sparse(I,J,s);
A=AG(UNK,UNK);
F=f(UNK)-AG(UNK,IB)*U(IB);
```

At this step, one needs a system of linear equations solver to find the solution of $AU = F$. Specifically:

- A direct method that uses a sparse Cholesky decomposition followed first by one forward substitution and next a backward substitution:

  ```
  M=chol(A);
  y=solvetril(M',F);% solves a lower triangular sparse system
  UC=solvetriu(M,y);% solves an upper triangular sparse system
  U(UNK)=UC;
  ```

- An iterative method that may use MATLAB command "pcg" which implements the pre-conditioned conjugate gradient method.

  ```
  %Without a pre-conditionner
  UC=pcg(A,B);
  U(UNK)=UC;
  ```

Theoretical Results for the One-Dimensional Discrete System

We prove now that A, is symmetric and positive definite (spd). We assume the following:

$$\forall x \in \overline{\Omega} : a(x) \geq \gamma_0, \ (x) \geq \gamma_1 \geq 0, \tag{7.18}$$

a, b, f are "sufficiently regular," specifically at least $C^k(\overline{\Omega})$, $k \geq 2$. (7.19)

Theorem 7.1 *Under (7.18), the matrix A in (7.15) has the following properties.*

1. *A is symmetric, diagonally dominant if $b_0 \geq 0$, strictly diagonally dominant if $b_0 > 0$.*

2. *A is symmetric positive definite.*

3. *$V^T A V \geq \gamma_0 \sum_{i=2}^{N-1} |\frac{V_{i+1}-V_i}{h}|^2 + \frac{V_1^2}{h2} + \frac{V_N^2}{h^2}$.*

4. *$||V||_2^2 \leq \frac{L^2}{\gamma_0} V^T A V$ and $||V||_\infty^2 \leq \frac{Lh}{\gamma_0} V^T A V$.*

Proof. We proceed successively:

1. Using (7.16) and (7.17), one has:

$$A_{P,i} + A_{O,i} + A_{E,i} = b_i \geq \gamma_1 \geq 0, \ i = 1, ..., N, \qquad (7.20)$$

and

$$\forall i, \ A_{E,i} = A_{O,i+1} = -\frac{1}{2h^2}(a_i + a_{i+1}) < 0, \text{ and } A_{P,i} > 0 \qquad (7.21)$$

which implies respectively diagonal dominance (strict when $\gamma_1 > 0$) and symmetry. ∎

2. When $\gamma_1 > 0$, strict diagonal dominance implies that A is spd.

Otherwise, consider without loss in generality the case when $b(x) = 0$, one then has: $V^T A V = V_1(A_{P,1}V_1 + A_{E,1}V_2) + ...$

$$.....+ \sum_{i=2}^{N-1} V_i(A_{P,i}V_i + A_{O,i}V_{i-1} + A_{E,i}V_{i+1}) + V_N(A_{P,N}V_N + A_{O,N}V_{N-1}).$$

Using (7.20), one has:

$$V^T A V = V_1(-(A_{O,1} + A_{E,1})V_1 + A_{E,1}V_2) + ...$$

$$..... + \sum_{i=2}^{N-1} V_i(-(A_{O,i} + A_{E,i})V_i + A_{O,i}V_{i-1} + A_{E,i}V_{i+1}) + ...$$

$$... + V_N(-(A_{O,N} + A_{E,N})V_N + A_{O,N}V_{N-1}).$$

Hence using (7.21), one has:

$$V^T A V = -(A_{O,1} + A_{O,2})V_1^2 + A_{O,2}V_1V_2 + ...$$

$$..... + \sum_{i=2}^{N-1} (-(A_{O,i} + A_{O,i+1})V_i^2 + A_{O,i}V_{i-1}V_i + A_{O,i+1}V_iV_{i+1}) + ...$$

$$... - (A_{O,N} + A_{E,N})V_N^2 + A_{O,N}V_{N-1}V_N.$$

Through an algebraic manipulation of the right-hand side of this last equality (left to Exercise 7.2), one obtains:

$$V^T A V = \begin{cases} -A_{O,1}V_1^2 + ... \\ ..... + \sum_{i=2}^{N} -A_{O,i}(V_{i-1}^2 - 2V_i V_{i-1} + V_i^2) + ... \\ ... - A_{E,N}V_N^2. \end{cases} \qquad (7.22)$$

Since, for all i $A_{O,i+1} = AE, i < 0$ and $|A_{O,i}| \geq \frac{\gamma_0}{h^2}$, this leads us to

$$V^T A V \geq \frac{\gamma_0}{h^2}(V_1^2 + \sum_{i=2}^{N}(V_{i-1} - V_i)^2 + V_N^2). \qquad (7.23)$$

Clearly $V^T A V = 0$ implies $V = 0$. ∎

3. The third part results from (7.23). ∎

4. The proof of the fourth part proceeds from the inequality:

$$\forall i \geq 2, \ V_i = V_1 + \sum_{j=2}^{i}(V_j - V_{j-1}) \text{ implying: } |V_i| \leq |V_1| + \sum_{j=2}^{i}|V_j - V_{j-1}|,$$

and hence $\forall i \geq 2$:

$$|V_i|^2 \leq i(|V_1|^2 + \sum_{j=2}^{i}|V_j - V_{j-1}|^2) \leq i(|V_1|^2 + \sum_{j=2}^{N}|V_j - V_{j-1}|^2 + |V_N|^2).$$

Therefore:

$$\sum_{i=1}^{N}|V_i|^2 \leq \frac{N(N+1)}{2}(|V_1|^2 + \sum_{j=2}^{N}|V_j - V_{j-1}|^2 + |V_N|^2).$$

Using (7.23), leads to:

$$\sum_{i=1}^{N}|V_i|^2 \leq N^2(|V_1|^2 + \sum_{j=2}^{N}|V_j - V_{j-1}|^2 + |V_N|^2) \leq \frac{L^2}{\gamma_0}V^T A V.$$

In a similar way we obtain the second estimate of part 4. ∎

As a consequence we prove existence and stability of the discrete system (7.15).

Theorem 7.2 (Existence and Stability) *Under the assumptions of Theorem 7.1, the discrete problem (7.15) has a unique solution that verifies:*

$$||U||_2 \leq \frac{L^2}{\gamma_0}||F||_2.$$

Proof. From Theorem 7.1-1, the matrix A is spd and therefore the system (7.15) has a unique solution. Furthermore, since:

$$U^T A U = U^T F,$$

then using part 4 of Theorem 7.1, one has:

$$||U||_2^2 \le \frac{L^2}{\gamma_0} U^T A U = \frac{L^2}{\gamma_0} U^T F \le \frac{L^2}{\gamma_0} ||U||_2 ||F||_2 \le \frac{L^2}{\gamma_0} ||U||_2 ||F||_2.$$

Simplifying by $||U||_2$ leads to the result of this theorem. ∎

Consequently, one obtains convergence of the discrete solution of (7.15) to the exact solution of (7.1) in addition to error estimates. Specifically if the solution u to (7.1) is such that, $u \in C^k(\Omega) \cap C(\overline{\Omega})$, $k \ge 3$, then the approximation $U_h = U = \{U_i\}$ to $u_h = \{u_i = u(x_i)\}$ verifies the estimates:

$$||u_h - U_h||_{N,2} \le c h^{k_0 - 2}, \ k_0 = \min\{k, 4\}, \tag{7.24}$$

where for $v \in \mathbb{R}^N$, $||v||_{N,2}^2 = \sum_{i=1}^{N} h |v_i|^2$.

The proof of this result is a classical procedure in numerical mathematics ([38]). It uses the two concepts of stability (Theorem 7.2) and consistency. This last concept is based on the estimation associated with the difference formula (7.11), namely if $u \in C^k$, $k \ge 3$, then:

$$u''(x_i) = \delta_h^2 u(x_i) + h^{k_0 - 2} \epsilon_i(u), \ v \in C^4, \ 1 \le i \le N, \ k_0 = \min\{k, 4\}.$$

where $\epsilon_i(u)$ depends on u and satisfies $||\epsilon||_\infty \le C$, with C independent from h. The completion of the proof (Exercise 7.3) is based on the equation $A(u - U) = h^{k_0 - 2} \epsilon$ with u considered here as the vector in $\mathbb{R}^{N+1}$ $\{u(x_i) \, | \, i = 0, ..., N\}$. ∎

7.3.2 Two-Dimensional Poisson-Dirichlet on a Rectangle

With $\Omega = (0, L) \times (0, l)$, $g : \Gamma \to \mathbb{R}$ and $a, f : \Omega \to \mathbb{R}$, consider the two-dimensional Poisson-Dirichlet problem:

$$\begin{cases} \mathcal{L}[u](x, y) = -\nabla(a(x, y)\nabla u(x, y)) = f(x, y), \forall \, (x, y) \in \Omega \\ u(x, y) = g(x, y), \, \forall \, (x, y) \in \Gamma = \Gamma_D, \end{cases} \tag{7.25}$$

We start by defining a uniform discrete domain on $\overline{\Omega}$.
Let Π_h and Π_k be respectively uniform partitions of the intervals $(0, L)$ and $(0, l)$:

$$\Pi_h = \{x_i = ih | 0 = x_0 < x_1 < .... < x_m = L\}, \ mh = L$$

and

$$\Pi_k = \{y_j = jh | 0 = y_0 < y_1 < .... < y_n = b\}, \ nk = l.$$

Consider then the discrete grids on Ω and $\overline{\Omega}$:

$$\Omega_{h,k} = \{P_{ij} = (x_i, y_j) | x_i \in \Pi_h, \, y_j \in \Pi_k, \, 0 < i < m, \, 0 < j < n.\}$$

$$\overline{\Omega_{h,k}} = \{P_{ij} = (x_i, y_j) | x_i \in \Pi_h, \, y_j \in \Pi_k, \, 0 \le i \le m, \, 0 \le j \le n.\}$$

and

$$\Gamma_{h,k} = \{P_{ij} = (x_i, y_j) | \, x_i = 0 \text{ or } y_j = 0 \text{ or } x_i = L \text{ or } y_j = l\}$$

Define then the discrete two-dimensional Poisson-Dirichlet system:

$$\begin{cases} \mathcal{L}_{h,k}[U]_{ij} = f_{ij}, \, (x_i, y_j) \in \Omega_{h,k}, \\ U_{ij} = g_{ij} = g(x_i, y_j), \, (x_i, y_j) \in \Gamma_{h,k}, \end{cases} \tag{7.26}$$

where, explicitly, one has:

$$\mathcal{L}_{h,k}[U]_{ij} = -\delta_h[a(x_i, y_j)\delta_h[U_{ij}]] - \delta_k[a(x_i, y_j)\delta_k[U_{i,j}]]$$

and therefore:

$$\mathcal{L}_{h,k}[U]_{ij} = \begin{cases} -\frac{1}{h^2} a(x_i + \frac{h}{2}, y_j)(u(x_i + h, y_j) - u(x_i, y_j))... \\ ... - \frac{1}{h^2} a(x_i - \frac{h}{2}, y_j)(u(x_i - h, y_j) - u(x_i, y_j))... \\ ... - \frac{1}{k^2} a(x_i, y_j + \frac{k}{2})(u(x_i, y_j + k) - u(x_i, y_j))... \\ ... - \frac{1}{k^2} a(x_i, y_j - \frac{k}{2})(u(x_i, y_j - k) - u(x_i, y_j)) \end{cases} \tag{7.27}$$

By regrouping the terms in (7.27), one gets:

$$\mathcal{L}_{h,k}[U]_{ij} = \begin{cases} \frac{1}{h^2}(a(x_i + \frac{h}{2}, y_j) + a(x_i - \frac{h}{2}, y_j))U_{i,j} \\ \frac{1}{k^2}(a(x_i, y_j + \frac{k}{2}) + a(x_i, y_j - \frac{k}{2}))U_{i,j} \\ ... - \frac{1}{h^2} a(x_i + \frac{h}{2}, y_j)U_{i+1,j} \\ ... - \frac{1}{h^2} a(x_i - \frac{h}{2}, y_j)U_{i-1,j} \\ ... - \frac{1}{k^2} a(x_i, y_j + \frac{k}{2})U_{i,j+1} \\ ... - \frac{1}{k^2} a(x_i, y_j - \frac{k}{2})U_{i,j-1} \end{cases} \tag{7.28}$$

The finite-difference discrete system corresponding to (7.7) is then defined as follows:
Find $U : \overline{\Omega}_{h,k} \longrightarrow \mathbb{R}$, such that:

$$\begin{cases} \mathcal{L}_{h,k}[U]_{ij} = f_{ij}, \forall (x_i, y_j) \in \Omega_{h,k}, \\ U_{ij} = g_{ij}, \forall (x_i, y_j) \in \Gamma_{h,k} = \partial\Omega_{h,k}, \end{cases} \tag{7.29}$$

with also: $f_{ij} = f(x_i, y_j)$ in $\Omega_{h,k}$ and $g_{ij} = g(x_i, y_j)$ on $\partial\Omega_{h,k}$.
Note when $a(x, y) = 1$,

$$\mathcal{L}_{h,k}[U]_{ij} = -\Delta_{h,k}[U]_{ij} := -(\delta_h^2[U_{ij}] + \delta_k^2[U_{ij}]), \, (x_i, y_j) \in \Omega_{h,k},$$

which is the well-known five-point difference scheme.

Implementation

We turn now to implementing (7.29). To put that system in matrix form:

$$AU = F, \tag{7.30}$$

$A \in \mathbb{R}^{N,N}$, $N = (m-1)(n-1)$, depends on how one indexes the points in $\overline{\Omega}_{h,k}$. As in the one-dimensional case, A is a sparse matrix with the vector $F \in \mathbb{R}^N$ taking into account the boundary conditions on U.

1. **Generating the grid**

   To each node $(\mathcal{S})$ of the grid $\Pi_{h,k}$ we associate the numerical values of its coordinates and also the values at $(\mathcal{S})$ of the functions a, f, g given in (7.7). One also needs to identify $(\mathcal{S})$ as to whether it is a boundary point, $\mathcal{S} \in \Gamma$ or an interior point, $\mathcal{S} \in \Omega$. Thus, for every point $(\mathcal{S})$, we define:

   - its coordinates $x((\mathcal{S}))$, $y((\mathcal{S}))$,

   - $i((\mathcal{S}))$ and $j((\mathcal{S}))$ the one-dimensional indices corresponding to the coordinates, $x((\mathcal{S})) = i((\mathcal{S})) \times h$, $y((\mathcal{S})) = j((\mathcal{S})) \times k$.

   - id indicates whether S is a boundary point or not: $id((\mathcal{S})) = 0$ means that $(\mathcal{S}) \in \Gamma_D$ and $id((\mathcal{S})) = 1$ identifies that $(\mathcal{S}) \in \Omega$.

   - When $(\mathcal{S})$ is an interior point, i.e., $id((\mathcal{S})) = 1$, one needs to define the corresponding values of a and f at $(\mathcal{S})$.

   - The attribute $U((\mathcal{S}))$ gives the discrete solution at $(\mathcal{S})$. If $(\mathcal{S})$ is a boundary point the value of g at $(\mathcal{S})$ is assigned for $U((\mathcal{S}))$. Otherwise for $id((\mathcal{S})) = 1$ we store the solution U of the discrete system (7.30).

   Thus the two-dimensional `Nodes` data structure is shaped up in Table 7.4.

I Node index not stored	xs	ys	is	js	id	a	f	US $(= g$ on $\Gamma_D)$ Discrete system solution
1								
N								

TABLE 7.4: Two-dimensional `Nodes` structure.

For this two-dimensional model, note that the total number of grid points in $\overline{\Omega}_{h,k}$ is $M = (m+1) \times (n+1)$ parameters and those in $\Omega_{h,k}$ number $N = (m-1) \times (n-1)$ corresponding to the unknowns components of U. Furthermore, the two-dimensional discrete domain is uniform if and only if:

$$|x_{i\pm1} - x_i| = h, \forall i \text{ and } |y_{j\pm1} - y_j| = k$$

and is globally uniform if $h = k$. In such case, the MATLAB function 7.4
implements the construction of the grid.

Algorithm 7.4 Generating a Two-Dimensional Nodes Structure

```
function [xs,ys,is,js,id,a,f,U]= Nodes2d(pih,pik)
% This function generates the table  Nodes for the 2d case
% INPUT: pih and pik partitions of [0,L] and [0,1],
% OUTPUT: Attributes of each node
% Generation of the table is done sequentially,
% by probing the grid horizontally or vertically
% I is the node index initialized to zero
I=0;
L=max(pih);m1=length(pih);%m1=m+1
l=max(pik);n1=length(pik);%n1=n+1
H=diff(pih);h=max(pih);
K=diff(pik);k=max(pik);
% if pih and pik are uniform, h and k is the mesh sizes
% of thes uniform partitions
    for j=1:n1
          for i=1:m1
              I=I+1;
              xs(I)=pih(i);ys(I)=pik(j);is(I)=i;yjs(I)=j;
              if i==1 |i==m |j==1 |j==n
    id(I)=0;
              else
                  id(I)=1;
              end
               %Assign values of other attributes
          end
    end
M=I;
```

On the other hand, Table 7.4 leads into determining the set of parameters that is necessary to implement the discrete system, in particular,
the number of unknowns N. For that purpose as in the one-dimensional
case, one uses the MATLAB command,

```
UNK=find(id);
% M is the number of unknowns in the system AU=B
N=length(UNK);
```

2. **Generating the coefficients of the linear system** $AU = F$
   Given that the grid on which the finite-differences discretization is
   based, is represented by the data structureNodes, then for each interior node, the discrete problem is defined using the 5-points difference

formula. That is, for each node $P = (i, j)$, with neighboring nodes: $W = (i - 1, j)$, $E = (i + 1, j)$, $S = (i, j - 1)$, $N = (i, j + 1)$, which we label respectively as "west," "east," "south" and "north" nodes. The uniform centered five-point finite difference formula is given at each node $S \in \Omega_{h,k}$, the equation of system (7.29) can be written as:

$$\forall (x_i, y_j) \in \Omega_{h,k} : \mathcal{L}_{h,k}[U]_{i,j} = a_P(S)U_{i,j} + \dots$$

$$a_O(S)U_{i-1,j} + a_E(S)U_{i+1,j} + a_S(S)U_{i,j-1} + a_N(S)U_{i,j+1}) = f_{i,j}. \quad (7.31)$$

Implementing the discrete system (7.31) requires the formulae for the coefficients $a_P(S)$, $a_O(S)$, $a_E(S)$, $a_S(S)$ and $a_N(S)$. These are obtained from (7.28) so as to give for every $S \in \Omega_{h,k}$:

$$\begin{cases} a_P(S) = \frac{1}{h^2}(a(x_i + \frac{h}{2}, y_j) + a(x_i - \frac{h}{2}, y_j)) + \dots \\ \dots + \frac{1}{k^2}(a(x_i, y_j + \frac{k}{2}) + a(x_i, y_j - \frac{k}{2})) \\ a_E(S) = -\frac{1}{h^2}a(x_i + \frac{h}{2}, y_j) \\ a_O(S) = -\frac{1}{h^2}a(x_i - \frac{h}{2}, y_j) \\ a_N(S) = -\frac{1}{k^2}a(x_i, y_j + \frac{k}{2}) \\ a_S(S) = -\frac{1}{k^2}a(x_i, y_j - \frac{k}{2}) \end{cases} \quad (7.32)$$

Note that for all $S \in \Omega_{h,k}$:

$$a_P(S) + a_E(S) + a_O(S) + a_N(S) + a_S(S) = 0.$$

For a square domain, with a globally uniform discretization under the parameters (h, k) with $h = k$, the values of the coefficients a, are given by:

$$a_S(S) = a_N(S) = a_O(S) = a_E(S) = -\frac{1}{h^2}, a_P(S) = \frac{4}{h^2}, \forall S \in \Omega_{h,k}.$$

We write now the procedures that generate the data structures associated with (7.31) using (7.32). These would allow solving the system (7.30).

As in the one-dimensional case, these data structures would allow utilizing the **sparse MATLAB** facilities. For that purpose, we generate a table COEFFA based on the model given in Table 7.2 which includes three columns:

(a) The first I is an interior node index that corresponds to an unknown value of the solution.

(b) The second J, $1 \leq J \leq N$, corresponds to all nodes in the table **Nodes** in a way where the coefficient s_{IJ} resulting from (7.32) is non-zero.

(c) The third column gives the value of this coefficient s_{IJ}.

nnz gives the number of non-zero coefficients. Algorithm 7.5 generates the non-zero coefficients of A.

Algorithm 7.5 Sparse Finite Difference Two-Dimensional Poisson Matrix

```
function [I,J,s,IB,UNK,f,U]=coeffa(pih,pik)
% UNK is the set of indices corresponding to the unknowns
% IB is the set of border indices
% J in IB: there exists an interior node such that s(I,J)~=0
% Pt is the coeffa table pointer. It is initialized to 0
UNK=[];IB=[];Pt=0;m=length(pih);n=length(pik);
[xs,ys,is,js,id,a,f,U]= Nodes(pih,pik);
%Computing the vector H={hi, i=1, ...,m-1}
% and K={kj, j=1, ...,n-1}
H=diff(pih);h=max(pih);K=diff(pik);h=max(pik);
for i=2:m-1
    for j=2:n-1%Then find ij node and its neighboring nodes
        Iij=find(is==i&js==j);UNK=[UNK Iij];
        IO=find(is==i-1&js==j);if i==2 IB=[IB IO];end
        IE=find(is==i+1&js==j);if i==m-1 IB=[IB IE];end
        IS=find(is==i&js==j-1);if j==2 IB=[IB IS];end
        IN=find(is==i&js==j+1);if j==n-1 IB=[IB IN];end
        %Computing the corresponding coefficients
        sO=-........;sE=-.........;
        sS=-..........;sN=-..........;
        sP=-(sO+sE+sS+sN);%Then fill the table coeffa
        Pt=Pt+1;I(Pt)=Iij;J(Pt)=Iij;a(Pt)=sP;
        Pt=Pt+1;I(Pt)=Iij;J(Pt)=IO;a(Pt)=sO;
        Pt=Pt+1;I(Pt)=Iij;J(Pt)=IE;a(Pt)=sE;
        Pt=Pt+1;I(Pt)=Iij;J(Pt)=IS;s(Pt)=sS;
        Pt=Pt+1;I(Pt)=Iij;J(Pt)=IN;s(Pt)=sN;
    end
end
```

7.3.3 Complexity for Direct Methods: Zero-Fill Phenomenon

To illustrate, assume the matrix $A \in \mathbb{R}^{N,N}$ in the profile of a banded structure:

$$
A = \begin{pmatrix}
a_{11} & a_{21} & \cdots & a_{1,\omega_2} & \cdots & \cdots & 0 \\
a_{21} & a_{22} & \cdots & a_{2,\omega_2+1} & \cdots & \cdots & 0 \\
\cdots & \cdots & \cdots & \cdots & \cdots & \cdots & \cdots \\
a_{\omega_1,1} & a_{\omega_1,2} & 0 & \cdots & \cdots & \cdots & 0 \\
0 & a_{\omega_1+1,2} & \cdots & \cdots & \cdots & \cdots & a_{n-\omega_2,n} \\
\cdots & \cdots & \cdots & \cdots & \cdots & \cdots & \cdots \\
0 & \cdots & 0 & \cdots & a_{n,n-\omega_1} & \cdots & a_{nn}
\end{pmatrix}
$$

We focus on solving $Ax = b$, with A a sparse spd matrix ($A = A^T$, $x^T Ax \geq 0$ if and only if $x \neq 0$), using Cholesky's decomposition for the matrix A, i.e., we seek the coefficients of the lower triangular matrix $L \in \mathbb{R}^{N,N}$ such that:

$$A = L \times L^T.$$

On the basis of Exercise 3.3, the total number of flops to obtain L for a full matrix A is $O(\frac{N^3}{3})$. However, considering the implementation of a Cholesky decomposition for a banded sparse matrix with a bandwidth of $2\omega + 1$, the decomposition algorithm becomes:

```
for j=1:N
    l(j,j)=sqrt(a(j,j));
    for i=j+1:min(j+omega,N)
        l(i,j)=a(i,j)/l(j,j);
    end
    for k=j+1:min(j+omega,N)
        for i=k:min(k+omega,N)
            a(i,k)=a(i,k)-l(i,j)*l{kj};
        end
    end
end
```

and the associated number of floating-point operations becomes:

- N square roots.
- $\sum_{j=1}^{N} \sum_{i=j+1}^{\min(N,j+\omega)} 1$ divisions $\leq N\omega$.
- $\sum_{j=1}^{N} \sum_{k=j+1}^{\min(N,j+\omega)} \sum_{i=k}^{\min(N,k+\omega)} 1$ multiplications and as many algebraic additions $\leq 2N(\omega^2 + \omega)$.

Thus, such decomposition requires a total number of flops of $O(2N\omega(\omega + 1))$ which is significantly less than $O(\frac{N^3}{3})$.

However, such decomposition is subject to the zero-fill phenomenon. Specifically, although A and L have the same bandwidth, Cholesky's decomposition and more generally any type of Gaussian decomposition transforms zeros within the matrix band of A into non-zero numbers within the band of L, thus leading into an inflation in the number of non-zero elements. Typically, if we are adopting a MATLAB type of matrix storage then $nnz(L) > nnz(A)$. We illustrate this phenomenon through the following example. Consider this segment of MATLAB program:

```
B=sparse(1:8,1:8,4)+sparse(2:8,1:7,-1,8,8)+sparse(1:7,2:8,-1,8,8);
I=speye(8);
A=[B -I;-I B];
U=chol(A);L=U';
subplot(1,2,1),spy(L)
subplot(1,2,2),spy(A)
```

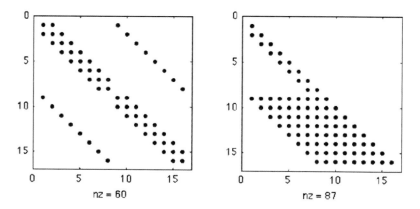

FIGURE 7.1: Non-zero fill: profiles of an spd matrix (left) and Cholesky's factor (right)

One then gets Figure 7.1.

Some error estimates

Similarly to the one-dimensional case, using the norm:

$$||v||_{N,2} = \{\sum_{i=1}^{m-1}\sum_{j=1}^{n-1} hk|v_{ij}|^2\}^{1/2} = (hkv^T v)^{1/2} = (hk)^{1/2}||v||_2.$$

one shows that:

$$||v||_{N,2} \leq \sqrt{Ll}||v||_\infty. \tag{7.33}$$

with furthermore A positive definite and

$$||v||_{N,2}^2 \leq \frac{L^2 + l^2}{2} hkv^T Av. \tag{7.34}$$

The proof proceeds from the inequalities:

$$|v_{i,j}|^2 \leq x_i h(\sum_{i=1}^{m-1} |\frac{v_{i+1,j} - v_{i,j}}{h}|^2) \leq Lh(\sum_{i=1}^{m} |\frac{v_{i+1,j} - v_{i,j}}{h}|^2), \tag{7.35}$$

and

$$|v_{i,j}|^2 \leq y_j k(\sum_{j=1}^{n-1} |\frac{v_{i,j+1} - v_{i,j}}{k}|^2) \leq lk(\sum_{j=1}^{n-1} |\frac{v_{i,j+1} - v_{i,j}}{k}|^2). \tag{7.36}$$

By multiplying inequality (7.35) with hk, then summing up over i and j yields:

$$\sum_{i=1}^{m-1}\sum_{j=1}^{n-1} hk|v_{i,j}|^2 \le a^2 hk \left(\sum_{j=1}^{n-1}\sum_{i=1}^{m-1} \left|\frac{v_{i+1,j} - v_{i,j}}{h}\right|^2\right).$$

Similarly, with inequality (7.36), one obtains:

$$\sum_{i=1}^{m-1}\sum_{j=1}^{n-1} hk|v_{i,j}|^2 \le b^2 hk \left(\sum_{i=1}^{m-1}\sum_{j=1}^{n-1} \left|\frac{v_{i,j+1} - v_{i,j}}{k}\right|^2\right).$$

Summing the last two inequalities leads to the proposed (7.34). ∎

Theorem 7.3 (Existence and Stability) *The discrete problem (7.30) has a unique solution that verifies:*

$$||U||_{N,2} \le \frac{L^2 + l^2}{2}||F||_{N,2}.$$

Proof. From Theorem 7.34-1, the matrix A is spd and therefore the system (7.30) has a unique solution. Furthermore, starting with:

$$||U||_{N,2}^2 \le \frac{a^2 + b^2}{2}hk U^T AU = \frac{a^2 + b^2}{2}U^T F \le \frac{a^2 + b^2}{2}||U||_{N,2}||F||_{N,2},$$

then simplifying by $||U||_{N,2}$ gives the proposed estimate. ∎

7.4 Variational Formulations

We turn now to finite-element type discretizations. These are formulated on the basis of variational formulations.

7.4.1 Integration by Parts and Green's Formula

Variational formulations are based on the use of the **integration by parts formula** in the one-dimensional calculus and on **Green's formula** in higher dimensions calculus.

In **one-dimensional** calculus, the integration by parts formula is well known and given by:

$$\int_0^L w'(x)v(x)dx = [w(x)v(x)]_0^L - \int_0^L v'(x)w(x)dx, \qquad (7.37)$$

$\forall\, v, w \,\in\, C^1((0, L)) \cap C([0, L])$. In particular, if $a \,\in\, C^1([0, L])$ and $u \,\in\, C^2((0, L)) \cap C^1([0, L])$, then by letting in (7.37) $w(x) = a(x)u'(x)$, one has:

$$\int_0^L (a(x)u'(x))'(x)v(x)dx = [a(x)u'(x)v(x)]_0^L - \int_0^L a(x)u'(x)v'(x)dx. \quad (7.38)$$

$\forall\, v \in C^1((0, L)) \cap C([0, L]),\ \forall\, u \in C^2((0, L)) \cap C^1([0, L])$.

In **multi-dimensional** calculus, Green's formula generalizes the integration by parts formula (7.37). In that case, the regularity of the underlining domain Ω is important in the formulation of Green's formula. In what follows, $\Omega \subset \mathbb{R}^d$ is assumed to be an open-bounded subset of class C^1 (or piecewise C^1), with boundary $\Gamma = \partial\Omega$. In two dimensions, a "smooth" boundary can be characterized as follows.

Definition 7.2 *Let Ω be an open-bounded domain of $\mathbb{R}^2$. Its boundary $\Gamma = \partial\Omega$ is considered to be smooth if it can be described by a finite sequence of vertices $\{P_i | i = 1, ..., k\}$, where $(P_1 = P_k)$ and correspondingly a sequence of simple arcs $\{\overrightarrow{P_i P_{i+1}} | i = 1, 2, ..., k-1\}$, whereby each arc $\overrightarrow{P_i P_{i+1}}$ is being described by a regular at least C^1 function $\psi_{i,i+1}(x, y) = 0$.*

There are several forms of Green's formula, of which we give the following:
Form 1 (Divergence form): $\forall\, V \in (C^1(\Omega) \cap C(\overline{\Omega}))^d,\ \forall\, v \in C^1(\Omega) \cap C(\overline{\Omega})$

$$\int_\Omega (div(V))v\, dxdy = \int_\Gamma (\nu.V)v\, ds - \int_\Omega (V.\nabla v)\, dxdy. \quad (7.39)$$

Several formulae result from (7.39).
Form 2 If in (7.39) we let $V = a(x)\nabla u$, where $u \in C^2(\overline{\Omega})$ and $a \in C^1(\Omega) \cap C(\overline{\Omega})$, then one obtains the second form of Green's formula,

$$\int_\Omega (div(a(x)\nabla u))v\, dxdy = \int_\Gamma (a(x)\nu.\nabla u)v\, ds - \int_\Omega (a(x)\nabla u.\nabla v)\, dxdy. \quad (7.40)$$

More particularly, if $V = \nabla u$ in (7.39), one obtains the Laplacian form:
Form 3 $\forall\, u \in C^2(\overline{\Omega}),\ \forall\, v \in C^1(\Omega) \cap C(\overline{\Omega})$

$$\int_\Omega (\Delta u)v\, dxdy = \int_\Gamma (\nu.\nabla u)v\, ds - \int_\Omega (\nabla u.\nabla v)\, dxdy. \quad (7.41)$$

Interestingly, variational formulations of boundary-value ordinary and partial differential equations are based on integration by parts and Green's formulae (7.38), (7.40) and (7.41).

7.4.2   Variational Formulation to One-Dimensional Poisson Problems

In the case of the one-dimensional model and if we let $\Omega = (0, L)$, the differential equation in (7.1) can be rewritten as:

$$-\frac{d}{dx}(a(x)\frac{du}{dx}) + b(x)u = f(x), x \in \Omega \, u(0) = \alpha, \, u'(L) = \beta. \qquad (7.42)$$

Although (7.42) can be handled using finite-difference discretizations, we find out now that it can be dealt with more naturally if it is put in variational form. We associate with this boundary value problem the two sets:

1. The set of "test functions":

$$\mathcal{T} = \{\varphi \in C^1(\Omega) \cap C(\overline{\Omega}) | \varphi(0) = 0\}$$

2. The set of "admissible functions":

$$U_{ad} = \{\varphi \in C^1(\Omega) \cap C(\overline{\Omega}) | \varphi(0) = \alpha.\}.$$

Note that $\mathcal{T}$ is a subspace of $C^1(\Omega) \cap C(\overline{\Omega})$ and U_{ad} is affine to $\mathcal{T}$ in the sense that if $u_0(x) = \alpha\frac{x}{L}$ then:

$$v \in U_{ad} \Leftrightarrow v = u_0 + \varphi, \, \varphi \in \mathcal{T}.$$

This is denoted by either $U_{ad} = \mathcal{T} + \{u_0\}$ or $U_{ad} = \mathcal{T} + \{\alpha\}$.
Let $\varphi \in \mathcal{T}$. Then, multiplying the differential equation in (7.1) by φ and integrating by parts yields:

$$\int_0^L (a(x)u'(x)\varphi'(x) + b(x)u(x)\varphi(x))dx = \int_0^L f(x)\varphi(x)dx + \beta\, a(L)\varphi(L), \, \varphi \in \mathcal{T}.$$

Define now:

$$A(v, \varphi) = \int_0^L (a(x)u'(x)\varphi'(x) + b(x)u(x)\varphi(x))dx$$

and

$$F(\varphi) = <f, \varphi> + \beta\, a(L)\varphi(L) = \int_0^L f(x)\varphi(x)dx + \beta\, a(L)\varphi(L).$$

Classical and variational formulations are related through the following:

Theorem 7.4 *Every solution u to (7.1) is a solution to:*

$$u \in U_{ad} = \mathcal{T} + \{\alpha\} : A(u, \varphi) = F(\varphi), \, \forall\, \varphi \in \mathcal{T}. \qquad (7.43)$$

7.4.3 Variational Formulations to Two-Dimensional Poisson Problems

Consider the following mixed Dirichlet-Neumann boundary value problem:

$$\begin{cases} \mathcal{L}[u](x,y) = -\nabla(a(x,y)\nabla u(x,y)) = f(x,y), \forall\,(x,y) \in \Omega \\ u(x,y) = g(x,y),\,\forall\,(x,y) \in \Gamma_D \subset \partial\Omega, \\ \nu.\nabla u(x,y) = h(x,y),\,\forall\,(x,y) \in \Gamma_N = \partial\Omega - \Gamma_D. \end{cases} \qquad (7.44)$$

where a and f are given functions on Ω and g, h and k are defined on $\partial\Omega$. As for the one-dimensional case, we associate with this boundary value problem the two sets:

1. The set of "test functions":

$$\mathcal{T} = \{\varphi \in C^1(\Omega) \cap C(\overline{\Omega}) | \varphi|_{\Gamma_D} = 0\}$$

2. The set of "admissible functions":

$$U_{ad} = \{\varphi \in C^1(\Omega) \cap C(\overline{\Omega}) | \varphi|_{\Gamma_D} = g\}.$$

$\mathcal{T}$ and U_{ad} have the same property as above, in the sense that there exists a function $u_0(x) \in C^1(\Omega) \cap C(\overline{\Omega})$ such that:

$$v \in U_{ad} \Leftrightarrow v = u_0 + \varphi,\; \varphi \in T.$$

One then writes $U_{ad} = \mathcal{T} + \{u_0\}$ and simply as:

$$U_{ad} = \mathcal{T} + \{g\}.$$

Let $\varphi \in \mathcal{T}$. Then, multiplying the differential equation in (7.7) by φ and using Green's formula (7.40) with $V = a(x,y)\nabla u$ yields:

$$\int_\Omega (a(x,y)\nabla u.\nabla\varphi + b(x,y)u.\varphi(x))dx\,dy \;=\; \int_\Omega f(x,y)\varphi(x,y)dx\,dy$$
$$+ \int_{\Gamma_N} h(s)\,ds,\,\varphi \in \mathcal{T}.$$

Define now:

$$A(v,\varphi) = \int_\Omega (a(x,y)\nabla u.\nabla\varphi + b(x,y)u\varphi)\,dxdy$$

and

$$F(\varphi) = \int_\Omega f(x,y)\varphi dxdy + \int_{\Gamma_N} h(s)\,ds,\,\varphi \in \mathcal{T}.$$

We have a similar variational formulation as that of the one-dimensional problem (7.43).

Theorem 7.5 *Every solution u to (7.7) is a solution to:*

$$u \in U_{ad} : A(u,\varphi) = F(\varphi),\,\forall\varphi \in \mathcal{T}. \qquad (7.45)$$

Conversely, it is easily verified that if $u \in U_{ad} \cap C^2(\Omega)$ and u is a solution to (7.45), then u solves also (7.7).

In the case of (7.6), this problem can be put into variational formulation, by considering:

1. The set of "test functions":

$$\mathcal{T} = \{\varphi \in C^1(\Omega) \cap C(\overline{\Omega}) | \varphi|_{\Gamma_D} = 0\}$$

2. The set of "admissible functions":

$$U_{ad} = \{\varphi \in C^1(\Omega) \cap C(\overline{\Omega}) | \varphi|_{\Gamma_D} = g\}.$$

One easily verifies the variational formulation:
$\forall v \in \mathcal{T}$,

$$u \in U_{ad} : \int_\Omega a(x,y)\nabla u.\nabla v \, dxdy = \int_\Omega f.v \, dxdy + \int_{\Gamma_N} h(s)v(s) \, ds, \quad (7.46)$$

If $A(u,v) = \int_\Omega a(x,y)\nabla u.\nabla v \, dxdy$ and $F(v) = \int_\Omega f(x,y).v(x,y) \, dxdy + \int_{\Gamma_N} h(s)v(s)ds$, then the variational formulation can be put in the form:

$$u \in U_{ad} : A(u,v) = F(v), \forall v \in \mathcal{T}. \quad (7.47)$$

In what follows C^1 spaces are replaced with spaces of PC^1 functions, whereas:

$$\varphi \in PC^1 \text{ if and only if } \varphi \in C \text{ and piecewise } C^1,$$

with:

$$\|\varphi\|_1^2 = <\varphi, \varphi>_1 = <\varphi, \varphi> + <\varphi', \varphi'>, \text{ in one dimension}$$

and:

$$\|\varphi\|_1^2 = <\varphi, \varphi>_1 = <\varphi, \varphi> + <\varphi_x, \varphi_x> + <\varphi_y, \varphi_y>, \text{ in two dimensions}$$

Similarly, C^2 spaces are replaced with PC^2 spaces of functions:

$$\varphi \in PC^2 \text{ if and only if } \varphi \in C^1 \text{ and piecewise } C^2.$$

In that manner, one defines PC^k functions for any k.

As a matter of fact, PC^1 spaces are not the "natural" spaces in which one may study existence of solutions of variational Equations (7.43), (7.45) and (7.47). Since these are defined in terms of bilinear forms $<.,.>$ and $A(.,.)$, it is then more appropriate to use "inner product spaces" H^1 of Sobolev spaces, discussed briefly in the Appendix to this chapter. However, in what follows, we limit ourselves to PC^1 with use of $\|\varphi\|_1^2 = <\varphi, \varphi>_1$.

More so, if $A(.,.)$ is a symmetric, continuous bilinear form on $PC^1 \times PC^1$, in addition to being coercive on $\mathcal{T} \times \mathcal{T}$, $(A(\varphi, \varphi) \geq c_0||\varphi||_2^2)$ and $F(.)$ being linear and continuous on PC^1, then using Lax-Milgram theorem, (7.47) admits a unique solution (see Appendix).

The user is invited to transform into variational forms the boundary value problem (7.1), with the respective boundary conditions: (7.2), (7.3), (7.4) and (7.5) in, respectively, Exercises 7.5, 7.6, 7.7 and 7.8.
Similarly, the same procedure would be applied to two-dimensional problems in Exercise 7.9.

7.4.4   Petrov-Galerkin Approximations

Given PC^1 with the scalar product: $< ., . >_1$ and consequent norm $||.||_1$, we consider the variational formulation:

$$u \in \mathcal{T} + \{g\} : A(u, v) = F(v), \forall\, v \in \mathcal{T}. \tag{7.48}$$

where $\mathcal{T}$ is a subspace of PC^1, g an element in PC^1.
The Petrov-Galerkin discretization proceeds as follows:

1. Select a set of linearly independent elements $\{\varphi_1, ..., \varphi_N\}$ in $\mathcal{T}$ such that $S_N = span\{\varphi_1, ..., \varphi_N\}$ "approximates" $\mathcal{T}$ in PC^1, i.e.,

$$\forall v \in \mathcal{T}, \exists v_N \in S_N : \lim_{N \to \infty} ||v - v_N||_1 = 0.$$

2. Let g_N be an approximation to g in PC^1, i.e., $\lim_{N \to \infty} ||g - g_N||_1 = 0$.

One then seeks

$$u_N \in S_N + \{g_N\} : A(u_N, v) = F(v), \forall\, v \in S_N. \tag{7.49}$$

Practically, one has:

$$u_N = g_N + \sum_{j=1}^{N} U_j \varphi_j.$$

Thus, one needs to find $\{U_j | j = 1, ..., N\}$ in order to find u_N. Note that (7.49) is equivalent to:

$$A(\sum_{j=1}^{N} U_j \varphi_j, \varphi_i) = F(\varphi_i) - A(g_N, \varphi_i), \forall\, i = 1, ..., N.$$

This in turn is equivalent to the system of linear equations:

$$A_N U = G_N, \tag{7.50}$$

where $A_N \in \mathbb{R}^{N,N} = \{A(\varphi_j, \varphi_i) | 1 \leq i, j \leq N\}$ and $G_N \in \mathbb{R}^N$, $G_{N,i} = F(\varphi_i) - A(g_N, \varphi_i)$. To implement (7.50), one needs to perform the following pre-processing tasks:

1. Generate the matrix $A_N = \{A(\varphi_i, \varphi_j)\}$

2. Generate $G_N : G_{N,j} = F(\varphi_j) - A(g_N, \varphi_j)$

3. Solve $A_N U = G_N$

On the basis of the positivity (coercivity) of the bilinear form $A(.,.)$ on $\mathcal{T} \times \mathcal{T}$ (and therefore on $S_N \times S_N$) and of its bi-continuity on $PC^1 \times PC^1$, in addition to the continuity of $F(.)$ on PC^1, one proves the following:

1. The system $A_N U = G_N$ admits a unique solution, with A_N positive definite and if $A(.,.)$ is symmetric, then A_N is symmetric positive definite.

2. Let $u = u^1 + g$, $u^1 \in \mathcal{T}$.
   If $\lim_{N \to \infty} \|g - g_N\|_1 = 0$ and if for every $v \in \mathcal{T}$, there exists $v_N \in S_N$ such that $\lim_{N \to \infty} \|v - v_N\|_1 = 0$, then there exits a constant C independent from N, such that:

   $$\|u - u_N\|_1 \leq C(\|g - g_N\|_1 + \|u^1 - v_N\|_1) \, \forall v_N \in S_N. \qquad (7.51)$$

   This implies $\lim_{N \to \infty} \|u - u_N\|_1 = 0$.

These are classical results that can be easily found in the literature (e.g., [41]).

7.5   One-Dimensional Finite-Element Discretizations

Consider the model problem (7.37), which variational formulation is given by:

$$u \in \mathcal{T} + \{a\} : A(u, v) = F(v), \, \forall \, v \in \mathcal{T}, \qquad (7.52)$$

with:

$$\mathcal{T} = \{\varphi \in PC^1(0, 1) \,|\, \varphi(0) = 0\}.$$

The assumptions on a, b, and f are such that:

1. All assumptions allowing bi-continuity, bi-linearity of $A(.,.)$ on $PC^1 \times PC^1$ and coercivity on $\mathcal{T} \times \mathcal{T}$ in addition to linearity and continuity for $F(.)$ on PC^1.

2. The unique solution u to (7.52) is at least in $PC^1(\Omega)$).

The solution $u(x)$ of (7.52) is such that:

$$J(u) = \frac{1}{2} A(u, u) - F(u) = \min_{v \in U_{ad}} \{J(v)\} \qquad (7.53)$$

Because $A(u, v) = A(v, u)$, note that (7.52) and (7.53) are equivalent.

7.5.1   The P_1 Finite-Element Spaces

1. Define first the "elements" on the domain Ω of the solution. In (7.42) $\Omega = (0,1)$, we let $\Omega : \bigcup_i [x_i, x_{i+1}]$ where $\{x_i \mid i = 1, ..., n\}$ is a set of nodes so that:

   (a) $x_i \neq x_j$ for $i \neq j$

   (b) $x_1 = 0$ and $x_n = 1$

   If E_i is $[x_i, x_{i+1}]$, then $\bigcup_i E_i = [0,1]$ and:

   $$E_i \cap E_j = \begin{cases} \phi \\ 1 \text{ Node} \\ E_i \text{ itself} \end{cases}$$

   Remark 7.1 *There is no uniformity on the elements, i.e., there exists i, j such that $x_{i+1} - x_i \neq x_{j+1} - x_j$.*

2. Define the P_1 finite-element spaces:

   $$S_1(\Pi) = \{\varphi \in C([0,1]) \mid \varphi|_{E_i} \text{ is a linear polynomial } \forall\, i\}.$$

We can now verify the following results:

Theorem 7.6 *Every $\varphi \in S_1(\Pi)$ is uniquely determined by its values at the nodes $\{P_i\}$ of the partition.*

Given $\{\varphi(x_i) \mid 1 \leq i \leq n\}$, then on $E_i = [x_i, x_{i+1}]$,

$$\varphi(x) = \varphi(x_i)\frac{x_{i+1} - x}{x_{i+1} - x_i} + \varphi(x_{i+1})\frac{x - x_i}{x_{i+1} - x_i}, \quad x \in [x_i, x_{i+1}] \qquad (7.54)$$

gives the expression of φ on $[x_i, x_{i+1}] \,\forall\, i$.

Definition 7.3 *For every node x_i, define*

$$i = 2...n-1 \quad \psi_i(x) = \begin{cases} \frac{x - x_{i-1}}{x_i - x_{i-1}} & x \in [x_{i-1}, x_i] \\ \frac{x_{i+1} - x}{x_{i+1} - x_i} & x \in [x_i, x_{i+1}] \\ 0 & \text{otherwise} \end{cases}.$$

$$i = 1 \qquad \psi_1(x) = \begin{cases} \frac{x_2 - x}{x_2 - x_1} & x \in [x_1, x_2] \\ 0 & \text{otherwise} \end{cases}$$

$$i = n \qquad \psi_n(x) = \begin{cases} \frac{x - x_{n-1}}{x_n - x_{n-1}} & x \in [x_{n-1}, x_n] \\ 0 & \text{otherwise} \end{cases}$$

Using (7.54), note that for $x \in [x_i, x_{i+1}]$, $\varphi(x) = \varphi(x_i)\psi_i(x) + \varphi(x_{i+1})\psi_{i+1}(x)$. This gives:

Theorem 7.7 $\forall \varphi \in S_1(\Pi), \varphi(x) = \sum_{i=1}^{n} \varphi(x_i)\psi_i(x)$, *i.e., $\{\psi_i(x)\}$ is a canonical Lagrangian basis for P_1 finite elements.*

Theorem 7.8 *Every $\varphi \in S_1(\Pi)$ is such that*

1. $\varphi \in C([0,1])$
2. $\varphi \in PC^1(0,1)$.

Proof. We proceed as follows:

Part 1 follows from the definition of φ.
For part 2, consider

$$\eta_i(x) = \begin{cases} \frac{1}{x_i - x_{i-1}} & x \in (x_{i-1}, x_i) \\ \frac{1}{x_i - x_{i+1}} & x \in (x_i, x_{i+1}) \end{cases}$$

It is easy to prove that η_i is the piecewise local derivative of ψ_i successively in (x_{i-1}, x_i) and (x_i, x_{i+1}).

Theorem 7.9 *If $v \in S_1(\Pi)$, then $v \in C([0,1]) \cap PC^1(0,1)$, with v' is piecewise constant.*

7.5.2 Finite-Element Approximation Using $S_1(\Pi)$

Let $S_{1,D}(\Pi) = \{v \in S_1(\Pi) | v(0) = 0\}$. The P_1 finite element approximation u_Π to the solution u of (7.52) is given by:

$$u_\Pi \in S_{1,D}(\Pi) + \{\alpha\} : A(u_\Pi, v) = F(v), \ \forall v \in S_{1,D}(\Pi). \tag{7.55}$$

$$J(u_\Pi) = \min_{v \in S_{1,D}(\Pi) + \{\alpha\}} \{J(v)\}. \tag{7.56}$$

Given $v \in S_{1,D}(\Pi)$, then

$$v(x) = v(x_2)\psi_2(x) + ... + v(x_n)\psi_n(x).$$

Also, if $v \in S_{1,D}(\Pi) + \{\alpha\}$, then $v(x) = \alpha\psi_1(x) + v(x_2)\psi_2(x) + ... + v(x_n)\psi_n(x)$. The FEM solution $u_\Pi(x) = \alpha\psi_1(x) + U_2\psi_2(x) + ... + U_n\psi_n(x)$ depends on $n-1$ unknowns.

$$u_\Pi(x) = \alpha\psi_1(x) + \sum_{i=2}^{n} U_i\psi_i(x),$$

with:

$$\begin{aligned} J(u_\Pi) &= \min_{v \in S_{1,D}(\Pi) + \{\alpha\}} \{J(v)\} \Longleftrightarrow u_\Pi \in S_{1,D}(\Pi) + \{\alpha\} : A(u_\Pi, v) \\ &= F(v), \ \forall v \in S_{1,D}(\Pi) \end{aligned}$$

Note that (7.55) is equivalent to:

$$A(u_\Pi, \psi_i) = F(\psi_i), \quad \forall i = 2 : n \tag{7.57}$$

because $S_{1,D}(\Pi) = \text{span}\{\psi_2, ..., \psi_n\}$.

Since $u_\Pi(x) = \alpha\psi_1(x) + \sum\limits_{j=2}^{n} U_j\psi_j(x)$, then (7.57) is equivalent to:

$$A(\sum_{j=2}^{n} U_j\psi_j(x), \psi_i) = F(\psi_i) - \alpha A(\psi_1, \psi_i), \quad \forall\, i = 2:n. \tag{7.58}$$

Since A is bilinear, then (7.58) is equivalent to:

$$\sum_{j=2}^{n}[A(\psi_j, \psi_i)]U_j = F(\psi_i) - \alpha A(\psi_1, \psi_i), \quad 2 \le i \le n. \tag{7.59}$$

(7.59) is equivalent to the system of linear equations:

$$KU = G \tag{7.60}$$

where $K \in \mathbb{R}^{n-1,n-1}$ called the **stiffness** matrix of the finite element system:
$U = \begin{pmatrix} U_2 \\ \vdots \\ U_n \end{pmatrix} \in \mathbb{R}^{n-1}$, and $G = \{F(\psi_i) - \alpha A(\psi_1, \psi_i) \,|\, i = 2...n\}$.

Theorem 7.10 *If $a(x) \ge a_0 > 0$ and $b(x) \ge 0$, then the matrix K is spd, i.e.,*
$$v^T K v \begin{cases} \ge 0 \,\forall\, v \in \mathbb{R}^{n-1} \\ = 0 \quad \text{iff } v = 0 \end{cases}$$
and therefore, the finite element system $KU = G$ has a unique solution.

Remark 7.2 *Generally, finite element stiffness matrices K are sparse. In fact, in one-dimensional problems if we use linear elements, then K is tridiagonal.*

This is verified as follows:

$$\begin{aligned} A(\psi_j, \psi_i) &= A(\psi_i, \psi_j) \\ &= <a(x)\frac{\mathrm{d}\psi_i}{\mathrm{d}x}, \frac{\mathrm{d}\psi_j}{\mathrm{d}x}> + <b(x)\psi_i, \psi_j> \\ &= A_1(\psi_i, \psi_j) + A_2(\psi_i, \psi_j) \end{aligned}$$

where $A_1(\psi_i, \psi_j) = \int\limits_0^1 a(x)\frac{\mathrm{d}\psi_i}{\mathrm{d}x}\frac{\mathrm{d}\psi_j}{\mathrm{d}x}\mathrm{d}x$ and $A_2(\psi_i, \psi_j) = \int\limits_0^1 a(x)\psi_i\psi_j\mathrm{d}x$.
Note that:

$$\frac{\mathrm{d}\psi_i}{\mathrm{d}x}(x) = 0, \,\forall\, x \notin (x_{i-1}, x_{i+1}) \text{ and } \frac{\mathrm{d}\psi_j}{\mathrm{d}x}(x) = 0, \,\forall\, x \notin (x_{j-1}, x_{j+1}).$$

Thus, $A_1(\psi_i, \psi_j) = 0$ if $|i - j| > 1$. A similar argument holds for $A_2(\psi_i, \psi_j)$. Hence, the matrix K shapes up as follows:

$$K = \begin{bmatrix} A(\psi_2, \psi_2) & A(\psi_2, \psi_3) & 0 & \cdots & \cdots \\ A(\psi_3, \psi_2) & A(\psi_3, \psi_3) & A(\psi_3, \psi_4) & 0 & \cdots \\ & & \ddots & & \\ 0 \cdots & A(\psi_i, \psi_{i-1}) & A(\psi_i, \psi_i) & A(\psi_i, \psi_{i+1}) & 0 \cdots \\ & & \ddots & & \\ 0 & \cdots & A(\psi_{n-1}, \psi_{n-2}) & A(\psi_{n-1}, \psi_{n-1}) & A(\psi_{n-1}, \psi_n) \\ 0 & \cdots & 0 & A(\psi_n, \psi_{n-1}) & A(\psi_n, \psi_n) \end{bmatrix}$$

7.5.3 Implementation of the Method

Implementing the method consists first in generating the elements $\{E_i \,|\, 1 \le i \le n-1\}$ which intimately depend on the nodes $\{x_i \,|\, 1 \le i \le n\}$. This generally requires two data structures, one for the `Nodes` and another for the `Elements`.

Structure `Nodes`

A node is specified by its geometrical type: interior node (1), boundary node (2) and its analytical type (3), indicating whether or not the value of the solution at node is known or unknown.
The former situation is the case for interior nodes or non-Dirichlet boundary nodes.
On that basis, we consider the following attributes:

$$T_G = \begin{cases} 0 & \text{interior points} \\ 1 & \text{boundary points} \end{cases}$$

$$T_U = \begin{cases} 0 & \text{unknown points: interior and non-Dirichlet boundary} \\ 1 & \text{known points: Dirichlet} \end{cases}$$

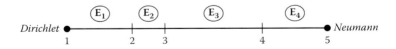

FIGURE 7.2: Layout for one-dimensional finite-element

X	Type		Parameters of the DE				U
	T_G	T_U	a	b	f	β	
x_1	1	1	-	-	-	0	α
x_2	0	0	-	-	-	0	-
x_3	0	0	-	-	-	0	-
x_4	0	0	-	-	-	0	-
x_5	1	0	-	-	-	β	-

TABLE 7.5: One-dimensional finite element Nodes structure.

Figure 7.2 and the consequent Table 7.5 illustrate the construction of the Nodes structure in this one-dimensional case.

Structure Elements

The basic attributes for this structure are shown in Table 7.6. To start with, we associate with each element, its vertices or end-point nodes.

Element Number	Valid	Boundary Points	
		i	j
1	1	1	2
2	1	2	3
3	1	3	4
4	1	4	5

TABLE 7.6: One-dimensional Elements structure for finite element.

Note that the attribute validity would be used for ulterior steps in case an element is split into two or more elements. In that case, the validity of the element being split becomes 0 and the added ones at the end of the table would have the validity 1. This is shown in Table 7.7.

Element Number	Valid	Boundary Points	
		i	j
1	0	1	2
2	1	2	3
3	1	3	4
4	1	4	5
5	1	1	1.5
6	1	1.5	2

TABLE 7.7: Edited one-dimensional Elements structure for finite element.

In what follows and for the sake of simplicity, we will omit the attribute "Validity" in the structure Elements.

Recall from (7.59) that the finite element system of linear equations is given by:

$$\sum_{j=2}^{n} [A(\psi_j, \psi_i)] U_j = F(\psi_i) - \alpha A(\psi_1, \psi_i), \quad 2 \le i \le n.$$

where

$$A(\psi_i, \psi_j) = \int_0^1 \left[a(x) \frac{\mathrm{d}\psi_i}{\mathrm{d}x} \frac{\mathrm{d}\psi_j}{\mathrm{d}x} + b(x)\psi_i \psi_j \right] \mathrm{d}x.$$

$\{\psi_i\}$ is a set of functions with compact support. $supp(\psi_i) = [x_{i-1}, x_{i+1}]$.

Definition 7.4 *Given a node i from the set of points that partition Ω we define:*

$$E(i) = \{ E \in \Pi \,|\, i \text{ is a boundary point to } E \}.$$

Example 7.1 *In the figure above, $E(3) = \{E_2, E_3\}, E(1) = \{E_1\}, E(5) = \{E_4\}$.*

Lemma 7.1

$$
\begin{aligned}
A(\psi_i, \psi_j) &= \int_{E(i) \cap E(j)} \left[a(x) \frac{\mathrm{d}\psi_i}{\mathrm{d}x} \frac{\mathrm{d}\psi_j}{\mathrm{d}x} + b(x)\psi_i \psi_j \right] \mathrm{d}x \\
&= \begin{cases} 0 & \textit{if } i \textit{ and } j \textit{ do not border the same element} \\ \ne 0 & \textit{if } i \textit{ and } j \textit{ border the same element} \end{cases}
\end{aligned}
$$

Corollary 7.1 *In the case of neighboring nodes (i and j border the same element) one has:*

1. $j = i$:

$$
\begin{aligned}
A(\psi_i, \psi_i) &= \int_{E_1} \left[a(x) \left(\frac{\mathrm{d}\psi_i}{\mathrm{d}x} \right)^2 + b(x)(\psi_i)^2 \right] \mathrm{d}x \\
&+ \int_{E_2} \left[a(x) \left(\frac{\mathrm{d}\psi_i}{\mathrm{d}x} \right)^2 + b(x)(\psi_i)^2 \right] \mathrm{d}x,
\end{aligned}
$$

   where $E(i) = \{E_1, E_2\}$.

2. $j \ne i$:

$$A(\psi_i, \psi_j) = \int_E \left[a(x) \frac{\mathrm{d}\psi_i}{\mathrm{d}x} \frac{\mathrm{d}\psi_j}{\mathrm{d}x} + b(x)\psi_i \psi_j \right] \mathrm{d}x,$$

   where E is the element with i and j as boundary points.

Thus, obtaining the matrix of the linear system (7.60) necessitates constructing new attributes for each element, specifically compute for each element local coefficients.

Local Stiffness Matrix

Define for an element E_{ij}, i.e., the element with boundary points i and j, the following matrix:

$$\begin{bmatrix} A_E(\psi_i,\psi_i) & A_E(\psi_i,\psi_j) \\ A_E(\psi_j,\psi_i) & A_E(\psi_j,\psi_j) \end{bmatrix}.$$

As a result,

$$A(\psi_i,\psi_j) = \begin{cases} i = j & A_{E_1}(i,i) + A_{E_2}(i,i) & E(i) = \{E_1,E_2\} \\ i \neq j & A_E(i,j) & E = E(i) \cap E(j) \end{cases}$$

This leads to completing Table 7.6 by adding these newly defined attributes. Excluding the attribute "Validity," the results of such addition are shown in Table 7.8.

El. (i_k,j_k)	A_E	F_E
1 (i_1,j_1)	$A_E(i_1,i_1)\ A_E(j_1,j_1)\ A_E(i_1,j_1)\ A_E(j_1,i_1)$	$(F_E(i_1),F_E(j_1))$
2 (i_2,j_2)	$A_E(i_2,i_2)\ A_E(j_2,j_2)\ A_E(i_2,j_2)\ A_E(j_2,i_2)$	$(F_E(i_2),F_E(j_2))$
$\vdots$	$\vdots$	$\vdots$
M (i_M,j_M)	$A_E(i_M,i_M)\ A_E(j_M,j_M)\ A_E(i_M,j_M)\ A_E(j_M,i_M)$	$(F_E(i_M),F_E(j_M))$

TABLE 7.8: A complete one-dimensional `Elements` structure for finite element.

Hence
$$A(\psi_i,\psi_j) = \ < a(x)\tfrac{d\psi_i}{dx},\tfrac{d\psi_j}{dx} > + < b(x)\psi_i,\psi_j >$$
$$= \sum_{E \in E(i) \cap E(j)} A_E(\psi_i,\psi_j)$$

Similarly,
$$F(\psi_i) = \ < f,\psi_i > + \beta a(1)\psi_i(1)$$
$$= \sum_{E \in E(i)} F_E(\psi_i)$$

where $\beta a(1)\psi_i(1) = \begin{cases} 0 & i \neq n \\ \beta a(1) & i = n \end{cases}$

Computing Local Matrix and Right-Hand Side Vector Elements

Local Matrix of Elements

$$A(\psi_i,\psi_j) = \int_E \left[a(x)\frac{d\psi_i}{dx}\frac{d\psi_j}{dx} + b(x)\psi_i\psi_j \right] dx$$

where $\psi_i\psi_j$ is a polynomial of degree 2 and $\frac{d\psi_i}{dx}\frac{d\psi_j}{dx}$ is a constant. Specifically if $E = E_{i,i+1} = (x_i, x_{i+1})$, then with $|E| = x_{i+1} - x_i$:

$$\psi_i(x) = \frac{x_{i+1} - x}{x_{i+1} - x_i}; \quad \frac{d\psi_i}{dx} = -\frac{1}{|E|},$$

$$\psi_{i+1}(x) = \frac{x - x_i}{x_{i+1} - x_i}; \quad \frac{d\psi_{i+1}}{dx} = \frac{1}{|E|}.$$

Thus:

$$\int_E [a(x)\frac{d\psi_i}{dx}\frac{d\psi_j}{dx}]dx = -\frac{1}{|E|^2}\int_E a(x)dx \simeq -\frac{1}{2|E|^2}[a_i + a_{i+1}]$$

with $|E|$ computed from the structure Nodes. On the other hand if:

$$\mu_{i,i+1} \equiv (\psi_i\psi_{i+1})(x) = \frac{1}{|E|^2}[(x_{i+1} - x)(x - x_i)]$$

then:

$$\int_E b(x)\mu_{i,i+1}(x)dx \simeq \frac{b(x_i) + b(x_{i+1})}{2}\int_{x_i}^{x_{i+1}} \mu_{i,i+1}(x)dx,$$

this approximation being done on the basis of the second mean value theorem:

$$\int_a^b f(x)w(x)dx = f(\xi)\int_a^b w(x)dx, \xi \in (a, b),$$

with $w(x) \geq 0$ and $f(x)$ both continuous.

Therefore one may use: $\int_a^b f(x)w(x)dx \approx \frac{f(a)+f(b)}{2}\int_a^b w(x)dx$.

Local Vector of Elements

Since $\psi_i(x) \geq 0$,

$$< f, \psi_i >_E = \int_{x_i}^{x_{i+1}} f(x)\psi_i(x)dx \simeq \frac{f(x_i) + f(x_{i+1})}{2}\int_{x_i}^{x_{i+1}} \psi_i(x)dx$$

Once the tables Nodes and Elements (based on Nodes), the coefficients of the local matrix and vector can then be computed using Algorithm 7.6.

Algorithm 7.6 Algorithm for Generating Local Data of Elements

```
for E = 1 : M
    if V == 1
        % V ==1 corresponds to element E being active
        % Compute local coefficients
        ...
    end
end
```

Assembling the Stiffness Matrix and Processing the Finite Element System

1. **Assembling** is implemented using:
$KG = \{A(\psi_i, \psi_j), \forall i, j\}$, the global stiffness matrix and $FG = \{F(\psi_i) \forall i\}$, the global vector through constructing two data structures as displayed in Table 7.9. For the purpose of assembling, the following MATLAB syntax can be used:

```
KG = sparse(I,J,AIJ);             % KG is an NxN matrix
FG = diag(sparse(I,I,FI));
IU = find (TU == 0);              % to find unknown  Nodes
IK = find (TU == 1);
K = KG(IU,IU);
G = FG(IU) - KG(IU,IK)*U(IK);
IN = find((TG == 1) & (TU == 0)); % to find the Neumann  Nodes
G(IN) = G(IN) + a(IN) .* beta(IN);
```

2. **Processing FE System**:
KG : Global stiffness matrix
FG : Global vector
IU : Index of unknown values of solution
IK : Index of known values of solution

Use some solver to find V that verifies $KV = F$: Direct Method (V = K\F) or iterative methods (i.e., Conjugate Gradient), then, insert V in the Nodes structure using the MATLAB instruction:
```
U(IU) = V;
```

I	J	AIJ
1	1	-
1	2	-
2	1	-
2	2	-
2	2	-
2	3	-
3	2	-
3	3	-

I	F(ψ_I)
1	-
2	-
2	-
3	-
3	-
4	-

TABLE 7.9: Local coefficients storage prior to using the `sparse` command.

7.5.4 One-Dimensional P_2 Finite-Elements

As for the P_1 case, we consider the nodes: $\{x_i \mid i = 1 : n\}$ and the elements: $\{E_j \mid j = 1 : m\}$. Let then:

$$S_2(\Pi) = \{\varphi \in C[0,1] \mid \varphi|_{E_i} \in \mathbb{P}_2, \forall\, i\}$$

To obtain a Lagrangian basis, we introduce the mid-points of the elements as in Figure 7.3. Thus, $\varphi \in S_2(\Pi)$ is totally defined by its values at the nodes and at the element mid-points (note that i_M is the middle node of element i).

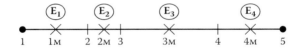

FIGURE 7.3: One-dimensional elements for P_2 elements

For $\varphi \in S_2(\Pi)$, one has for $x \in [x_i, x_{i+1}]$,

$$
\begin{aligned}
\varphi(x) &= \varphi_i \frac{(x_{i+1} - x)(x_{i+\frac{1}{2}} - x)}{\frac{h_i^2}{2}} + \varphi_{i+\frac{1}{2}} \frac{(x - x_i)(x_{i+1} - x)}{\frac{h_i^2}{4}} \\
&\quad + \varphi_{i+1} \frac{(x - x_i)(x - x_{i+\frac{1}{2}})}{\frac{h_i^2}{2}}
\end{aligned}
$$

Let then for each interior node i, $2 \leq i \leq n-1$:

$$\psi_i(x) = \begin{cases} \dfrac{(x-x_{i+1})(x-x_{i+\frac{1}{2}})}{\frac{h_i^2}{2}} & x \in [x_i, x_{i+1}] \\[3ex] \dfrac{(x-x_{i-1})(x-x_{i-\frac{1}{2}})}{\frac{h_{i-1}^2}{2}} & x \in [x_{i-1}, x_i] \\[3ex] 0 & \text{otherwise} \end{cases}$$

and at the boundary nodes:

$$\psi_1(x) = \begin{cases} \dfrac{(x-x_2)(x-x_{\frac{3}{2}})}{\frac{h_1^2}{2}} & x \in [x_1, x_2] \\[3ex] 0 & \text{otherwise} \end{cases}$$

$$\psi_n(x) = \begin{cases} \dfrac{(x-x_{n-1})(x-x_{n-\frac{1}{2}})}{\frac{h_{n-1}^2}{2}} & x \in [x_{n-1}, x_n] \\[3ex] 0 & \text{otherwise} \end{cases}$$

Note that $\psi_i(x_i) = 1$ and $\psi_i(x_j) = 0, \forall\, j \neq i$.

A Lagrangian basis of $S_2(\Pi)$ consists of:

$$\{\psi_i(x) \,|\, i = 1 : n\} \cup \{\eta_i(x) \,|\, i = 1 : m\}$$

where:

$$\eta_i(x) = \begin{cases} \dfrac{(x-x_i)(x_{i+1}-x)}{\frac{h_i^2}{4}} & x_i \leq x \leq x_{i+1} \\[3ex] 0 & \text{otherwise} \end{cases}$$

Obviously, one has:

$$\forall \varphi \in S_2(\Pi), \varphi(x) = \sum_{i=1}^{n} \varphi_i \psi_i(x) + \sum_{i=1}^{m} \varphi_{i+\frac{1}{2}} \eta(x).$$

Implementation

The finite element solution $u_\Pi(x) = \sum_{j=1}^{n} U_j \psi_j(x) + \sum_{j=1}^{m} U_{j+\frac{1}{2}} \eta_j(x)$ satisfies:

$$A(u_\Pi(x), v) = F(v) \quad \forall\, v \in S_2(\Pi),$$

i.e.,

$$A\left(\sum_{j=1}^{n} U_j \psi_j(x) + \sum_{j=1}^{m} U_{j+\frac{1}{2}} \eta_j(x), v\right) = F(v).$$

Hence for $v = \psi_i, \quad i = 1...n$ and $v = \eta_i, \quad i = 1...m$, one has:

$$\sum_{j=1}^{n} A(\psi_j, \psi_i) U_j + \sum_{j=1}^{m} A(\eta_j, \psi_i) U_{j+\frac{1}{2}} = F(\psi_i),$$

$$\sum_{j=1}^{n} A(\psi_j, \eta_i) U_j + \sum_{j=1}^{m} A(\eta_j, \eta_i) U_{j+\frac{1}{2}} = F(\eta_i),$$

which is equivalent to the system:

$$KU = F.$$

As in the P_1 case, to obtain the matrix K and the vector F, one needs to construct the equivalent of Tables 7.8 and 7.9 for storing local coefficients of matrix K and vector F, specifically:

$$\{A_E(\psi_i, \psi_i),\ A_E(\psi_i, \psi_{i+1}),\ A_E(\psi_i, \eta_i),\ A_E(\psi_{i+1}, \psi_i),\ A_E(\psi_{i+1}, \psi_{i+1})\},$$

$$\{A_E(\psi_{i+1}, \eta_i),\ A_E(\eta_i, \psi_i),\ A_E(\eta_i, \psi_{i+1}),\ A_E(\eta_i, \eta_i)\},$$

and

$$\{F(\psi_i),\ F(\psi_{i+1}),\ F(\eta_i).\}$$

Implementation of the structures Nodes, Elements, KG, FG (and consequently K and F) is done in a similar way to the P_1 case.

7.6 Exercises

Exercise 7.1

Perform a Taylor's series analysis to show that:

$$(a(x)v'(x))' = \delta_h(a(x)\delta_h v(x)) + O(h^2), \ v \in C^4.$$

What regularity should we impose on $a(x)$?

Exercise 7.2

Prove equality (7.22).

Exercise 7.3

Complete the proof of (7.24).

Exercise 7.4

Give a complete proof of (7.34).

Exercise 7.5

Put in variational form the boundary value problem (7.1) with (7.2). Verify some or all the assumptions on $A(.,.)$ on $F(.)$.

Exercise 7.6

Put in variational form the boundary value problem (7.1) with (7.3). Verify some or all the assumptions on $A(.,.)$ on $F(.)$.

Exercise 7.7

Put in variational form the boundary value problem (7.1) with (7.4). Verify some or all the assumptions on $A(.,.)$ on $F(.)$.

Exercise 7.8

Put in variational form the boundary value problem (7.1) with (7.5). Verify some or all the assumptions on $A(.,.)$ on $F(.)$.

Exercise 7.9

Put in variational form the boundary value problem (7.9). Verify some or all the assumptions on $A(.,.)$ on $F(.)$.

7.7 Computer Exercises

Computer Exercise 7.1 *Consider the implementation of the system resulting from (7.14):*

$$-\delta_h[a(x_i)\delta_h][U_i]] + b_iU_i = f_i = f(x_i), \; x_i \in \Omega_h, \; U_0 = \alpha, \; U_N = \beta,$$

with for $\Omega = (0, L)$, $a(x) = 1 + x^2$, $b(x) = \sin^2(x)$, with a-priori given solution $u(x) = (1 + x)\sin(x)$.
α, β, $f(x)$ are consequently obtained from $u(0)$, $u(1)$ and $f(x) = \mathcal{L}[u](x) = -\frac{d}{dx}(a(x)\frac{du}{dx}) + b(x)u$.
For that purpose, you are requested to perform the following tasks:

1. Generate Table 7.1.

2. Generate Table 7.2.

3. Therefore, the system $AU = F$ is generated using Algorithm 7.3 and the system can be solved accordingly using a direct method:

```
M=chol(A);
y=solvetril(M',F);% solves a lower triangular sparse system
UC=solvetriu(M,y);% solves an upper triangular sparse system
U(UNK)=UC;
```

Write the MATLAB functions and perform tests for:

$$h = 1/4, \; 1/8, \; 1/10, \; 1/16, \; 1/20.$$

Give your results in a table form:

Test results for one-dimensional finite difference approximations to one-dimensional Poisson equation on the unit interval				
h	cond(A)	$\max_i |u_i - U_i|$	$\frac{\max_i |u_i - U_i|}{h^2}$	Elapsed time
1/4				
1/8				
1/10				
1/16				
1/20				

4. For $h = 1/10, \; 1/16, \; 1/20$, plot on the same graph:

$$\{u(x_i), \; U_i \,|\, x_i \in \overline{\Omega_h}\}.$$

Computer Exercise 7.2 *Consider the implementation of the system:*

$$\begin{cases} \mathcal{L}_{h,k}[U]_{ij} = f_{ij}, \; (x_i, y_j) \in \Omega_{h,k}, \\ U_{ij} = g_{ij} = g(x_i, y_j), \; (x_i, y_j) \in \Gamma_{h,k}, \end{cases}$$

with $\mathcal{L}_{h,k}$ corresponding to the Laplace operator $-\Delta = -\frac{\partial^2}{\partial x^2} - \frac{\partial^2}{\partial y^2}$

\multicolumn{5}{c}{Results for the case $-\Delta u = xy(1-x)(1-y)$ in Ω}

| \multicolumn{5}{c}{$u = 0$ on $\partial\Omega$} |

| \multicolumn{5}{c}{considering the case $h = 1/32$ to replace u as its closest} |

h	N	nnz	$\frac{nnz}{N}$	$\|U_h - U_{1/32}\|_\infty$
$1/4$				
$1/8$				
$1/16$				

For $h = 1/4$, $1/8$, $1/16$, $1/32$, plot for each case on a two-dimensional graph:

$$\{U_{ij} \,|\, (x_i, y_j) \in \overline{\Omega_h}\}$$

Computer Exercise 7.3 *Similar to Computer Exercise 7.2, perform now a comparative study to compare the direct solver of (MATLAB: \\) with the gradient iterative solvers indicated in the table that follows.*

\multicolumn{6}{c}{Comparative study of elapsed times}
h	Direct	St. Des. (no PC)	CG no PC	Pre-cond CG M Inc. Chol	condest $(M^{-1} * A)$
$1/4$					
$1/8$					
$1/16$					
$1/32$					

Computer Exercise 7.4

Implementation and Testing of 1D P_1 Finite Element Method Consider implementing a one-dimensional P_1 finite element method on the following mixed boundary value problem:

$$-\frac{d}{dx}[a(x)\frac{du}{dx}] + b(x)u(x) = f(x),\ 0 < x < 1; \qquad (7.61)$$

with the boundary conditions:

$$u(0) = \alpha,\ u'(1) + u(1) = \beta,\ \epsilon \geq 0,$$

and

$$a(x) = 1 + x^2,\ b(x) = 1 - x^2,\ f(x) = x^2,\ \alpha = \beta = 1.$$

Implement P_1 finite element method to solve (7.61), using any type of mesh (uniform or non-uniform). Your software should generate the data structures: Nodes, Elements, sparse global stiffness matrix,...

Solve the system of linear equations using the MATLAB command: \\ and save your solutions in order to test for convergence.

Conduct tests on uniform meshes: $(n-1)h = 1$, $h = \frac{1}{4}, \frac{1}{8}, \frac{1}{10}, \frac{1}{16}$. Test convergence of the method to verify $O(h^2)$ convergence in L^∞ by considering the solution obtained from $h = \frac{1}{16}$ as the exact solution.

Appendix: Variational Formulations in Sobolev Spaces

For more details the reader is referred to [13], Chapter 5, p. 131.

Hilbert spaces. Orthonormal bases. Parseval's equality
Let H be a vector space, on which one defines a scalar product (u, v), i.e., a bilinear form from $H \times H \to \mathbb{R}$, that is symmetric and positive definite, i.e.,

$$(i)\,(u, v) = (v, u) \text{ and } (ii)\,(u, u) \geq 0, \text{ with } (u, u) > 0 \text{ iff } u \neq 0.$$

The scalar product induces a norm on H:

$$||u|| = ||u||_H = (u, u)^{1/2},$$

that verifies:

$$|(u, v)| \leq ||u||.||v|| \text{ (Schwarz inequality)} \qquad (7.62)$$
$$||u + v|| \leq ||u|| + ||v|| \qquad (7.63)$$

It also verifies the "parallelogram" identity:

$$||\frac{u + v}{2}||^2 + ||\frac{u - v}{2}||^2 = \frac{1}{2}(||u||^2 + ||v||^2), \forall\, u, v \in H.$$

H becomes then a **pre-Hilbert space** that is the basis of a Hilbert space.

Definition 7.5 *H is said to be a **Hilbert space**, if every Cauchy sequence $\{u_n\} \in H$ ($\lim_{m,n \to \infty} ||u_m - u_n|| = 0$) is convergent in H (i.e., H is a complete normed space).*

Remark 7.1 *Note that a complete normed vector space V is a **Banach space**. It is only when the norm of V is induced by a scalar product on V, that H becomes a Hilbert space.*

Basic Hilbert space: $L^2(\Omega)$, $\Omega \subset \mathbb{R}^d$, $d == 1, 2, ..$, where the scalar product is given by:

$$< u, v >= \int_\Omega u.v dx (dx = dx_1 dx_2...dx_d),$$

with the subsequent norm:

$$||u|| = ||u||_{0,2} = ||u||_{L^2(\Omega)} = \{\int_\Omega u^2 dx\}^{1/2}.$$

Note that $\{L^p(\Omega)|p \neq 2,\, 1 \leq p \leq \infty\}$ are Banach spaces, with norm:

$$||u||_{0,p} = ||u||_{L^p(\Omega)} = \{\int_\Omega u^p dx\}^{1/p}.$$

Definition 7.6 *A Hilbert (or Banach) space is said to be separable, if there exists a dense, infinitely countable basis of H.*

Hilbert spaces generalize the notion of finite dimensional Euclidean spaces. This is illustrated in the statement of the following theorem.

Theorem 7.11 *Every separable Hilbert space H, admits a complete (infinitely countable) orthonormal basis, $E = \{e_n | n = 1, 2, ...\}$, i.e.,:*

1. *$||e_n|| = 1$, $\forall n$, and $(e_i, e_j) = 0, i \neq j$.*

2. *The vector space generated by E is dense in H.*

3. *$\forall u \in H$, $u = \Sigma_{n=1}^{\infty}(u, e_n)e_n$, and*

4. *$||u||^2 = \Sigma_{n=1}^{\infty}|(u, e_n)|^2$ (Parseval's equality).*

5. *To every l^2 sequence $\{\alpha_n\}$ in $\mathbb{R}$, $\Sigma_{n=1}^{\infty}|\alpha_n|^2 < \infty$, $u_N = \Sigma_{n=1}^{N}\alpha_n e_n$ converges to an element $u \in H$, such that $\alpha_n = (u, e_n), \forall n$. (This fact proves isometry between any separable Hilbert space and the space l^2). Furthermore:*

$$||u - u_N||^2 = \Sigma_{n=N+1}^{\infty}|\alpha_n|^2 \qquad (7.64)$$

Sobolev spaces H^k, $W^{k,p}$

For more details, the reader is referred to [13], Chapter 8 p. 201. L^2-spaces (and more generally L^p-spaces, $1 \leq p < \infty$) are crucial in defining weak formulations of partial differential equations. These normally use the concept "Lebesgue integral" which generalizes "Riemann Integral." To circumvent Lebesgue integral, we present L^2 Elements as limit points of "regular" functions. We start by defining weak derivatives of L^p functions. For that purpose, we need a "practical" set of flexible test functions.

Definition 7.7 *$C_0^{\infty}(\Omega)$ is the space of functions that are infinitely differentiable, with* **compact support** *in Ω.*

Such space contains more than the zero function. Indeed, the support of a function ϕ defined in some open domain Ω of $\mathbb{R}^d$ is the closure of the set of points of Ω for which ϕ is different from zero. Thus, one can say that a function of compact support in Ω is a function defined on Ω such that its support Λ is a closed bounded set located at a distance from the boundary Γ of Ω by a number $\delta > 0$, where δ is sufficiently small. One considers infinitely differentiable functions of compact support through the function ψ, defined on the unit-sphere:

$$\psi(x) = \begin{cases} \exp\left(-1/(|x|^2 - 1)\right) & |x| < 1 \\ 0 & |x| \geq 1 \end{cases}$$

This function can be scaled to serve as an example of infinitely differentiable functions with compact support in any open domain Ω that can have a sphere as a subset. We may now give a modified statement of a result from [13] p. 61.

Theorem 7.12 *To every L^p-Cauchy sequence $\{\varphi_n\} \subset C_0^\infty(\Omega)$ (i.e., $\lim_{m\to\infty, n\to\infty} ||\varphi_m - \varphi_n||_{L^p(\Omega)} = 0$) has its limit point $\varphi \in L^p(\Omega)$.*

The limit point φ is such that $\lim_{n\to\infty} \int_\Omega |\varphi - \varphi_n|^p = 0$ and is unique as a function on Ω up to a set of "zero measure." As a result, the set $L^p(\Omega)$ is complete. It is also separable in the sense that there exists a dense countably infinite subset: $S = \text{span}\{\psi_1, \psi_2, ..., \psi_k, ...\}$, such that $\overline{S} = H$. For L^p spaces, this is well-stated in the following:

Theorem 7.13 *For $1 \le p < \infty$, there exists an **infinitely countable** basis S of $C_0^\infty(\Omega)$ that is dense in $L^p(\Omega)$, i.e., $\forall\, \varphi \in L^p(\Omega)$, there exists a sequence $\{\varphi_n\}$ in S, such that*

$$\lim_{n\to\infty} ||\varphi_n - \varphi||_{L^p(\Omega)} = 0.$$

Let

$$||\varphi||_p = ||\varphi||_{L^p(\Omega)} = \left(\int_\Omega |\varphi|^p dx\right)^{1/p}.$$

When $p = 2$,

$$||\varphi|| = ||\varphi||_2 = <\varphi, \varphi>^{1/2} = \left(\int_\Omega |\varphi|^2 dx\right)^{1/2}.$$

Assume the space-dimension is $d = 1$ and consider an L^p Cauchy sequence in $C_0^1(0, L)$, whereby

$$\lim_{m,n\to\infty} ||\varphi_n - \varphi_m||_p = 0 \quad \text{and} \quad \lim_{m,n\to\infty} ||\varphi_n' - \varphi_m'||_p = 0.$$

Then there exists a unique pair $\{\varphi, \psi\} \in (L^p(\Omega))^2$, such that:

$$\lim_{n\to\infty} ||\varphi_n - \varphi|| = 0 \quad \text{and} \quad \lim_{n\to\infty} ||\psi_n - \psi|| = 0.$$

One verifies that:

$$<\varphi, v'> = - <\varphi', v>, \forall\, v \in C_0^\infty(0, L).$$

This is done through a limit process as:

$$<\varphi, v'> = \lim_{n\to\infty} <\varphi_n, v'> = - \lim_{n\to\infty} <\varphi_n', v> = - <\psi, v>.$$

We will say that ψ is the weak derivative of ϕ as in the following definition.

Definition 7.8 *Let $\Omega \subseteq \mathbb{R}^d$, $u \in L^p(\Omega)$. The function $g \in L^p(\Omega)$ is said to be the **weak derivative** of u with respect to x_i, $i = 1, ...d$, if:*

$$\int_\Omega u \frac{\partial\phi}{\partial x_i} = -\int_\Omega g\phi, \forall\, \phi \in C_0^\infty(\Omega).$$

When $v \in L^p$ has a weak derivative with respect to x_i, it is unique and denoted by:

$$\frac{\partial v}{\partial x_i} - \text{weak.}$$

On the basis of weak derivatives, one introduces Sobolev spaces.

Definition 7.9 *Let $p \in \mathbb{R}$, $1 \le p \le \infty$.*

$$W^{1,p}(\Omega) = \{v \in L^p(\Omega) | \frac{\partial \phi}{\partial x_i} - \text{weak, } \forall\, i = 1, ...d\}$$

Introducing on $W^{1,p}(\Omega)$, the norm:

$$||v||_{1,p} = \{||v||_{L^p}^p + \Sigma_{i=1}^p ||v_{x_i}||_{L^p}^p\}^{1/p}, \ 1 \le p < \infty. \qquad (7.65)$$

For $p = \infty$, one uses the norm

$$||v||_{1,\infty} = \{||v||_{L^\infty} + \Sigma_{i=1}^p ||v_{x_i}||_{L^\infty}.$$

When $p = 2$, one deals with $H^1(\Omega)$, which norm results from the scalar product:

$$< u, v >_1 = < u, v > + \Sigma_{i=1}^d < \frac{\partial u}{\partial x_i}, \frac{\partial v}{\partial x_i} >,$$

in the sense that:

$$||u||_{1,2} \equiv ||u||_1 = \{< u, u >\}^{1/2}.$$

Proposition 7.1 *$H^1(\Omega)$ is a separable Hilbert space. Also, $W^{1,p}(\Omega)$ is a separable Banach space for $1 \le p < \infty$.*

We give now a weak version of the trace theorem for functions in $H^1(\Omega)$ or in $W^{1,p}(\Omega)$.

Theorem 7.14 *Trace of functions in Sobolev Spaces ([13] p. 315) If a function $u \in H^1(\Omega)$ ($u \in W^{1,p}(\Omega)$), with Ω an open bounded subset of $\mathbb{R}^d$, with a regular boundary $\Gamma = \partial\Omega$ sufficiently smooth (piecewise C^1), then:*

$$u|_\Gamma \in L^2(\Gamma), \ (u|_\Gamma \in L^p(\Gamma)).$$

Furthermore the trace operator $T : H^1(\Omega) \to L^2(\Omega)$ is bounded, in the sense that:

$$||u_\Gamma||_{L^2(\Gamma)} \le C(||u||_{H^1(\Omega)}),$$

Thus, one may define subspaces of $H^1(\Omega)$, namely:

$$H_0^1(\Omega) = \{v \in H^1(\Omega) | v = 0, \text{ on } \Gamma\},$$

and

$$H_D^1(\Omega) = \{v \in H^1(\Omega) | v = 0, \text{ on } \Gamma_D \subseteq \Gamma\},$$

Remark 7.2 *The trace theorem can be extended to functions that are $H^2(\Omega)$ and more generally in $H^m(\Omega)$. Specifically:*

$$\text{If } u \in H^2(\Omega), \text{ then } u|_\Gamma \in H^1(\Gamma), \nabla u \in (L^2(\Gamma))^d,$$

with the trace operator being bounded from $H^2(\Omega)$ to $H^1(\Gamma)$.

Sobolev Embeddings Results
Let $\Omega \subset \mathbb{R}^d$. Then, ([13] pp. 212-278):

1. $d = 1$:

   (a) $H^1(\Omega) \subset C(\overline{\Omega})$,

   (b) More generally: $H^m(\Omega) \subset C^{m-1}(\overline{\Omega})$.

2. $d = 2$:

   (a) $H^1(\Omega) \subset L^p(\Omega), \ \forall\, p \in [1, \infty)$.

   (b) For $p > 2$, $W^{1,p}(\Omega) \subset C(\overline{\Omega})$.

   (c) As a result, if $u \in H^2(\Omega) \subset C(\overline{\Omega})$.

Poincaré's Inequality In special cases, this inequality provides equivalence between H^1 and gradient seminorms defined by:

$$v \in H^1(\Omega) : |v|_1 = \left(\int_\Omega |\nabla v|^2 dx \right)^{1/2}.$$

Theorem 7.15 *For $v \in H_0^1(\Omega)$ and more generally for*

$$v \in H_D^1(\Omega) = \{v \in H^1(\Omega) | v = 0 \text{ on } \Gamma_D \subseteq \Gamma\},$$

there exists a constant $0 < c < 1$, such that:

$$c||v||_1 \le |v|_1 \le ||v||_1.$$

The proof is given in some special cases, when Ω is one-dimensional and rectangular two-dimensional problems with 0 boundary condition on one side of the rectangle. ∎

Weak formulations in Sobolev spaces. Lax-Milgram Theorem We start first by reformulating (7.46) and (7.47) on the just defined Sobolev spaces. Let:

$$U_{ad} = \{\varphi \in H^1(\Omega) | \varphi = g \text{ on } \Gamma_D\}.$$

We assume that g is sufficiently regular, i.e., piecewise continuous on Γ_D, so that it can be extended to a function u_0 in $H^1(\Omega)$, i.e.,

$$u_0 \in H^1(\Omega) : u_0 = g \text{ on } \Gamma_D.$$

Such extension is made possible, using the Calderon-Zygmund theory, which proves such extension with the property:

$$||u_0||_{1,\Omega} \leq K||g||_{\infty,\Gamma_D}.$$

Let:

$$H_D^1(\Omega) = \{\varphi \in H^1(\Omega)|\varphi = 0 \text{ on } \Gamma_D\}.$$

Thus, one has:

$$U_{ad} = u_0 + H_D^1(\Omega).$$

(simply written as $g + H_D^1(\Omega)$). Note that the expressions in (7.46) are all well defined, specifically in:

$$u \in U_{ad} : \int_\Omega a(x,y)\nabla u.\nabla v dxdy = \int_\Omega f(x,y)v(x,y)dxdy + ...$$

$$... + \int_{\Gamma_N} h(s)v(s)ds, \forall\, v \in H_D^1(\Omega),$$

one requires the assumptions: $f \in L^2(\Omega)$, $a \in C(\overline{\Omega})$, $h \in L^2(\Gamma_N)$. If $A(u,v) = \int_\Omega a(x,y)\nabla u.\nabla v dxdy$ and $F(v) = \int_\Omega f(x,y)v(x,y)dxdy + \int_{\Gamma_N} h(s)v(s)ds$, then the equivalent variational form is given by:

$$u \in \{u_0\} + H_D^1(\Omega) : A(u,v) = F(v), \forall\, v \in H_D^1(\Omega).$$

Lax-Milgram Theorem We state now the Lax-Milgram theorem, whereby:

H is a Hilbert space, $(H^1(\Omega))$ and V a closed subspace of $H(H_D^1(\Omega))$.

Let $A : H \times H \to \mathbb{R}$ be a bilinear form that is:

1. Bi-continuous on H: $|A(v,w)| \leq c_1||v||.||w||$, $\forall\, v,w \in H$.

2. Coercive on V: $A(v,v) \geq c_0||v||^2$, $\forall\, v \in V$.

Let also $F : H \to \mathbb{R}$ be a continuous linear operator such that $|F(v)| = L||v||$, $\forall\, v \in H$. Then:

1. $\exists$ a unique $u \in \{u_0\} + V$ verifying:

$$A(u,v) = F(v), \forall\, v \in V. \tag{7.66}$$

Furthermore:

2. If $A(.,.,)$ is also symmetric on H, then the unique solution u to (7.66) minimizes the energy $J(v) = \frac{1}{2}(A(v,v) - F(v))$, i.e.,

$$J(u) = \min_{v \in \{u_0\}+V} J(v). \tag{7.67}$$

Proof.

1. Uniqueness follows from coercivity of $A(.,.)$ on V.

2. Existence in the general case requires preliminary results:

   (a) The first of these results consists in the validity of Lax-Milgram when $A(.,.) = (.,.)_H$, the scalar product on H. In that case, one uses Riesz representation theorem to obtain the existence of an element $f \in H$ such that:

   $$F(v) = (f, v)_H, \; \forall \, v \in H.$$

   Thus if $\pi_V(f - u_0)$ is the orthogonal projection of $f - u_0$ on V, then $u = u_0 + \pi_V(f - u_0)$, verifies:

   $$(u, v)_H = F(v), \forall \, v \in V.$$

   (b) The second result is obtained from the bi-continuity of $A(.,.)$ and Riesz representation. It allows the existence of a linear continuous operator $L : H \to H$, such that:

   $$A(z, v) = (Lz, v)_H, \; \forall \, z, v \in H.$$

   with $||Lz||_H \leq C||z||_H, \forall \, z \in H$.

   To complete the proof, we consider for $\rho > 0$, the map $T_\rho : \{u_0\} + V \to \{u_0\} + V$, whereby:

   $$z = T_\rho(w)$$

   is given by

   $$(z, v)_H = (w, v)_H + \rho(A(w, v) - F(v)).$$

   From the above two steps, note that:

   $$z = T_\rho(w) = w - \rho(Lw - f).$$

   Let $z_i = T_\rho(w_i)$, $i = 1, 2$. Then one obtains:

   $$||z_1 - z_2||_H^2 = \rho^2||L(w_1 - w_2)||_H^2 - 2\rho A(w_1 - w_2, w_1 - w_2) + ||w_1 - w_2||_H^2.$$

   Using the coercivity of $A(.,.)$ on V and the continuity of L on H, one has:

   $$||z_1 - z_2||_H^2 \leq (1 - (2\rho c_0 - \rho^2 C^2))||w_1 - w_2||_H^2.$$

   Obviously for any choice of ρ such that $2\rho c_0 - \rho^2 C^2 > 0$, i.e., $0 < \rho < \frac{2c_0}{C^2}$, there exists a constant $0 < \kappa_\rho < 1$, such that the operator T_ρ is contractive, specifically for $\rho = \rho_0 = \frac{c_0}{C^2}$, $\kappa_0 = \sqrt{1 - \frac{c_0^2}{C^2}}$ and:

   $$||z_1 - z_2||_H \leq \kappa_0||w_1 - w_2||.$$

Using the Banach fixed point theorem, there exists a $u \in \{u_0\} + V$, such that:

$$u = T_\rho(u).$$

Such u verifies $A(u, v) = F(v), \forall v \in V$. Furthermore, the iteration: $\{u^{(n)} \in u_0 + V | n \geq 0\}$,

$$(u^{(n)}, v)_H = (u^{(n-1)}, v)_H - \rho_0(A(u^{(n-1)}, v) - F(v)), \forall v \in V.$$

$u^{(0)} = u_0$, converges to the unique solution of $A(u, v) = F(v)$ with:

$$||u^{(n)} - u||_H \leq \kappa_0^n ||u^{(0)} - u||.$$

3. Symmetric case. We first show that $(7.66) \Rightarrow (7.67)$.
   For that purpose, we show that $J(v)$ has a minimum m such that $\exists u \in \{u_0\} + V : J(u) = m$ and $A(u, v) = F(v), \forall v \in V$. Note that:

$$J(v) = \frac{1}{2}(A(v, v) - F(v)) \geq \frac{c}{2}(||v||^2 - L||v||) = g(||v||)$$

by coercivity of A and continuity of F. Hence $J(v) \geq m_0, \forall v \in H$, leading to:

$$\exists m \geq m_0 : m = min_{v \in H}[J(v)].$$

Hence, there exists $\{v_n\}$ which is minimizing for J, i.e.,

$$lim_{n \to \infty} J(v_n) = m.$$

Therefore:

$$\forall \epsilon > 0, \exists n_0 : n > n_0, \text{ and } m \leq J(v_n) < m + \epsilon.$$

This implies that $J(||v_n||) < m + \epsilon$ and $\exists M$ such that $||v_n|| \leq M, \forall n$. One concludes the existence of a subsequence $\{v_{n_i}\}$ of $\{v_n\}$ and an element $u \in \{u_0\} + V$ such that, by continuity, as $n_i \to \infty$, $J(v_{n_i}) \to J(u)$. By renaming $v_{n_i} = v_n$, we show now that:

$$m = J(u).$$

This is done as follows:

$$
\begin{aligned}
J(v_n) - J(u) &= \frac{1}{2}(A(v_n, v_n) - A(u, u)) - F(v_n - u) \\
&= \frac{1}{2}(A(v_n - u, v_n + u)) - F(v_n - u).
\end{aligned}
$$

Thus:

$$J(v_n) - J(u) = \frac{1}{2}(A(v_n - u, v_n - u) + A(u, v_n - u)) - F(v_n - u),$$

which is equivalent to:

$$J(v_n) - J(u) = \frac{1}{2}A(v_n - u, v_n - u) + 0 \geq 0.$$

As $n \to \infty$, we have:

$$m - J(u) \geq 0,$$

i.e., $m \geq J(u)$, and $m \leq J(u) \leq m \Rightarrow J(u) = m$.
The converse is obvious.

                                                        ■

Bibliography

[1] *MUMPS: A parallel sparse direct solver.* http://graal.ens-lyon.fr/MUMPS/.

[2] *SuperLU (Supernodal LU).* http://crd-legacy.lbl.gov/ xiaoye/SuperLU/.

[3] *UMFPACK: unsymmetric multifrontal sparse LU factorization package.* http://www.cise.ufl.edu/research/sparse/umfpack/.

[4] W. F. Ames. *Numerical Methods for Partial Differential Equations.* Academic Press, 1977.

[5] E. Anderson, Z. Bai, C. Bischof, S. Blackford, J. Demmel, J. Dongarra, J. Du Croz, A. Greenbaum, S. Hammerling, A. McKenney, and D. Sorensen. *LAPACK Users' Guide.* SIAM, Philadelphia, third edition, 1999.

[6] E. Andersson and P-A. Ekström. Investigating googles pagerank algorithm, 2004.

[7] D. Austin. How google finds your needle in the web's haystack.

[8] S. Bancroft. An algebraic solution of the GPS equation. *IEEE Trans. Aerospace and Electronic Systems*, AES-21:56–59, 1985.

[9] M. Benzi. Preconditioning techniques for large linear systems: A survey. *Computational Physics*, page 182, 2002.

[10] Å. Björck. Solving linear least squares problems by Gram-Schmidt orthogonalization. *BIT*, 7:1–21, 1967.

[11] R. F. Boisvert, R. Pozo, K. Remington, R. Barrett, and J. Dongarra. The Matrix Market: A Web repository for test matrix data. In R.F. Boisvert, editor, *The Quality of Numerical Software, Assessment and Enhancement*, pages 125–137. Chapman & Hall, London, 1997.

[12] C. Brezinski, M. Redivo Zaglia, and H. Sadok. New look-ahead Lanczos-type algorithms for linear systems. *Numerische Mathematik*, 83:53–85, 1999.

[13] H. Brezis. *Functional Analysis, Sobolev Spaces and Partial Differential Equations* . Springer, 2010.

[14] S. Brin and L. Page. The anatomy of a large-scale hypertextual web search engine. *Appl. Math. Letters*, pages 107–117, 1998.

[15] A.M. Bruaset. *A survey of preconditioned iterative methods*. Pitman Research Notes in Mathematics Series. Longman Scientific and Technical, 1995.

[16] P. A. Businger and G. H. Golub. Linear least squares solutions by Householder transformations. *Numer. Math.*, 7:269–276, 1965.

[17] J. Chaskalovic. *Finite Elements Methods for Engineering Sciences*. Springer, 2008.

[18] F. Chatelin, M. Ahues, and W. Lederman. *Eigenvalues of Matrices: Revised Edition*. Classics in Applied Mathematics. Society for Industrial and Applied Mathematics, 2012.

[19] P. Ciarlet. *Analyse Numérique Matricielle et Optimisation*. Masson, Paris, 1985.

[20] P.G. Ciarlet. *Introduction to Numerical Linear Algebra and Optimisation*. Cambridge University Press, Cambridge, 1989.

[21] J. K. Cullum and R. A. Willoughby. *Lanczos algorithms for large symmetric eigenvalue computations*. SIAM, Philadelphia, 2002.

[22] T.A. Davis. Algorithm 915, SuiteSparseQR: multifrontal multithreaded rank-revealing sparse QR factorization. *ACM Trans. Math. Softw.*, 38(1):8:1–8:22, 2011.

[23] J. Demmel, I. Dhillon, and H. Ren. On the correctness of some bisection-like parallel eigenvalue algorithms in floating point arithmetic. *Electron. Trans. Num. Anal. (ETNA)*, 3:116–149, 1995.

[24] J.J. Dongarra, J.J. Du Croz, I.S. Duff, and S.J. Hammarling. A set of level 3 basic linear algebra subprograms. *ACM Transactions on Mathematical Software*, 16:1–17, 1990.

[25] J.J. Dongarra, J.J. Du Croz, S.J. Hammarling, and R. J. Hanson. An extended set of Fortran Basic Linear Algebra Subprograms. Technical memorandum 41, Argonne National Laboratory, September 1986.

[26] J.J. Dongarra, I.S. Duff, D.C. Sorensen, and H. van der Vorst. *Numerical Linear Algebra for High-Performance Computers*. SIAM, 1998.

[27] J. Erhel. *Computational Technology Reviews*, volume 3, chapter Some Properties of Krylov Projection Methods for Large Linear Systems, pages 41–70. Saxe-Coburg Publications, 2011.

[28] J. Erhel. Erreur de Calcul des ordinateurs, IRISA Publications, 1990.

[29] R. Fletcher. Conjugate gradient methods for indefinite systems. *Lecture Notes Math*, pages 73–89, 1976.

[30] R. Freund, M. Gutknecht, and N. Nachtigal. An implementation of the look-ahead Lanczos algorithm for non-hermitian matrices. *SIAM journal on scientific computing*, 14:137–158, 1993.

[31] K. Gallivan, W. Jalby, and U. Meier. The use of BLAS3 in linear algebra on a parallel processor with a hierarchical memory. *SIAM J. Sci. Statist. Comput.*, 8(6):1079–1084, Nov. 1987.

[32] F.R. Gantmacher. *The Theory of Matrices*, volume 1. Chelsea Pub. Co., New York, N.Y., 2nd edition, 1959.

[33] G.H. Golub and C.F. Van Loan. *Matrix Computations*. The Johns Hopkins University Press, Baltimore, 3d edition, 1996.

[34] F. B. Hildebrand. *Finite-Difference Equations and Simulations*. Prentice Hall, 1968.

[35] A. S. Householder. *The Theory of Matrices in Numerical Analysis*. Dover Pub., New York, 1964.

[36] IEEE. Standard for floating-point arithmetic. *IEEE Std 754-2008*, pages 1–70, 2008.

[37] S. Inderjit, Dhillon, Beresford, and N. Parlett. Orthogonal eigenvectors and relative gaps. *SIAM J. Matrix Anal. Appl*, page 25, 2004.

[38] E. Isaacson and H.B. Keller. *Analysis of Numerical Methods, 4th edition*. John Wiley & Sons, Inc. New York, NY, USA, 1966.

[39] A. Iserles. *A First Course in the Numerical Analysis of Differential Equations, 2nd edition*. Cambridge University Press, Cambridge, UK, 2009.

[40] W. Jalby and B. Philippe. Stability analysis and improvement of the block Gram-Schmidt algorithm. *SIAM J. Sci. Stat. Comput.*, 12(5):1058–1073, Sep. 1991.

[41] C. Johnson. *Numerical Solution of Partial Differential Equations by the Finite Element Method*. Cambridge University Press, 1987.

[42] C.T. Kelley. *Iterative Methods for Linear and NonLinear Equations*. SIAM, USA, 1995.

[43] C. Lanczos. Solution of systems of linear equations by minimized iterations. *Journal of Research*, 49:1–53, 1952.

[44] C. Lawson, R.J. Hanson, D.R. Kincaid, and F.T. Krogh. Basic linear algebra subprograms for Fortran usage. *ACM Transactions on Mathematical Software*, 5(3):308–323, 1979.

[45] R. Lehoucq, D.C. Sorensen, and C. Yang. *ARPACK User's Guide: Solution of Large-Scale Eigenvalue Problems With Implicitly Restarted Arnoldi Methods*. SIAM, Philadelphia, 1998.

[46] R.B. Lehoucq and D.C. Sorensen. Deflation techniques for an implicitly restarted Arnoldi iteration. *SIAM J. Matrix Anal. Appl.*, 17:789–821, 1996.

[47] G. Meurant. *Computer solution of large linear systems*. North Holland, Amsterdam, 1999.

[48] D. Nuentsa Wakam and J. Erhel. Parallelism and robustness in GMRES with the Newton basis and the deflated restarting. *Electronic Transactions on Numerical Analysis*, 40:381–406, 2013.

[49] C. Paige and M. Saunders. Solution of sparse indefinite systems of linear equations. *SIAM journal on numerical analysis*, 12:617–629, 1975.

[50] C. C. Paige. Practical use of the symmetric Lanczos process with re-orthogonalization. *BIT*, 10:183–195, 1971.

[51] C. C. Paige. Computational variants of the Lanczos method for the eigenproblem. *J. Inst. Math. Appl.*, 10:373–381, 1972.

[52] C.C. Paige. *The computation of eigenvalues and eigenvectors of very large sparse matrices*. PhD thesis, London University, London, England, 1971.

[53] C.C. Paige and D. S. Scott. The Lanczos algorithm with selective or-thogonalization. *Math. Comp.*, 33:217–238, 1979.

[54] B. Parlett, D. Taylor, and Z. Liu. A look-ahead Lanczos algorithm for unsymmetric matrices. *Mathematics of computation*, 44:105–124, 1985.

[55] B.N. Parlett. *The Symmetric Eigenvalue Problem*. SIAM, 1998.

[56] David Poole. *Linear Algebra: A Modern Introduction (with CD-ROM)*. Brooks Cole, 2005.

[57] Y. Saad. *Numerical Methods for Large Eigenvalue Problems*. Halstead Press, New York, 1992.

[58] Y. Saad. *Iterative Methods for Sparse Linear Systems, 2nd edition*. SIAM, Philadelphia, PA, 2003.

[59] Y. Saad and M. H. Schultz. GMRES: A generalized minimal residual algorithm for solving nonsymmetric linear systems. *SIAM J. Sci. Statist. Comput.*, 7(3):856–869, July 1986.

[60] Scilab. Software available at http:/www.scilab.org and documentation at http:/wiki.scilab.org.

[61] V. Simoncini and D. Szyld. recent computational developments in Krylov subpsace methods for linear systems. *Numerical linear algebra with applications*, 14:1–59, 2007.

[62] G.L. Sleijpen and D.R. Fokkema. Bicgstab(l) for linear equations involving unsymmetric matrices with complex spectrum. *Electronic Transactions on Numerical Analysis*, 1:11–32, 1993.

[63] P. Sonnefeld. CGS, a fast Lanczos-type solver for nonsymmetric linear systems. *SIAM journal on scientific and statistical computing*, 10(36-52), 1989.

[64] G.W. Stewart and J.G. Sun. *Matrix Perturbation Theory*. Academic Press, Boston, 1990.

[65] G. Strang. *Introduction to Linear Algebra*. Wellesley-Cambridge Press, 1993.

[66] G. Strang and G. Fix. *An Analysis of The Finite Element Method*. Prentice-Hall, New Jersey, 1973.

[67] J.M. Tang, S.P. MacLachlan, R. Nabben, and C. Vuik. A comparison of two-level preconditioners based on multigrid and deflation. *SIAM. Journal on Matrix Analysis and Applications*, 31:1715–1739, 2010.

[68] J.M. Tang, R. Nabben, C. Vuik, and Y.A. Erlangga. Comparison of two-level preconditioners derived from deflation, domain decomposition and multigrid methods. *Journal of Scientific Computing*, 39:340–370, 2009.

[69] L.N. Trefethen and D. Bau, III. *Numerical Linear Algebra*. SIAM, Philadelphia, 1997.

[70] H.A. Van Der Vorst. Bi-CGSTAB: a fast and smoothly converging variant of Bi-CG for the solution of nonsymmetric linear systems. *J. Sci. Stat. Comput.*, pages 631–644, 1992.

[71] C.F. Van Loan. *Introduction to Scientific Computing: A Matrix-Vector Approach Using MATLAB*. Prentice-Hall, New Jersey, 1996.

[72] O.C. Zienkiewicz, R.L. Taylor, and J.Z. (Sixth ed.) Zhu. *The Finite Element Method: Its Basis and Fundamentals*. Butterworth-Heinemann, 2005.

Index